ION TRANSPORT IN HEART

To Professor Silvio Weidmann
on the occasion of his 60th birthday
April 1981

Ion Transport in Heart

L. J. Mullins, Ph.D.
Professor of Biophysics and Chairman
Department of Biophysics
University of Maryland School of Medicine
Baltimore, Maryland

Raven Press ▪ New York

Raven Press, 1140 Avenue of the Americas, New York, New York 10036

Made in the United States of America

Library of Congress Cataloging in Publication Data

Mullins, L. J. (Lorin John), 1917–
Ion transport in heart.

Includes bibliographical references and index.
1. Heart–Muscle. 2. Ions. 3. Biological transport.
4. Electrophysiology. I. Title. (DNLM: 1. Myocardium–Metabolism. 2. Ion exchange. 3. Electrophysiology.
WG 280 M959i)
QP133.2.M84 599.01'16 80-6272
ISBN 0-89004-645-X AACR2

Preface

In working out the details of Na/Ca exchange in squid axons, it seemed reasonable to ask: Can these principles be applied to other cells? More particularly, is there a cell with a very long duration action potential so that its electrochemical gradient for Na is greatly decreased for a significant period of time?

The answer to this question led directly to the development of this volume, because the reply was yes: the cardiac cell has all the properties one would expect if a significant control of contraction resided in a carrier-mediated rather than a channel-mediated process. Moreover, not only Ca entry but also its exit from the cell could be brought about by a single entity: Na/Ca exchange. This fact provides the rationale for the approach of this volume.

Because this book attempts to look at one particular phase of cardiac physiology—transport—the papers cited are those that I analyze in detail to illustrate a particular point. My purpose is not to review the field critically or to deny the existence of other, contrary views. It is general experience that the membrane-bound entities we call the Na/K pump, the Na channel, or Na/Ca exchange are universally distributed. Hence, this volume suggests that the long duration of the cardiac action potential has been developed for one and only one purpose: to reduce the Na electrochemical gradient for an appreciable period of time.

The treatment of the subject is largely nonmathematical and has been designed for students and research workers with a knowledge of cardiac electrophysiology, but without experience in ion transport across membranes. Chapters 1 through 4 provide some background in transport useful for these readers. Coverage of all chapters overlaps so that each chapter can be read and used individually.

L. J. Mullins
April 1981

Acknowledgments

I have had the advice and encouragement of many friends during the writing of this volume. In particular, Silvio Weidmann and Glenn Langer were kind enough to comment in detail on an early version of the manuscript; Paul Cranefield and David Gadsby have made comments on a more recent draft. For all this work on the part of others, I am most grateful.

This book was completed during my stay at the Instituto Venezolano de Investigaciones Cientificas, Caracas, Venezuela. I greatly appreciate the hospitality offered.

Contents

Chapter 1

Introduction

Studies of cardiac electrophysiology have been concerned with an identification of the currents that flow during a voltage clamp. These follow the classic studies of Hodgkin and Huxley, who used an analysis of voltage clamp currents to specify the patterns of conductance changed to both Na and K that underlie the generation of an action potential in nerve. In cardiac muscle, however, the recognition that there is a slow inward channel (Ca current) in addition to Na and K channels and the suggestion that K movement is fractionated among a variety of different types of K channels have led to great difficulty in analyzing voltage clamp data in classic Hodgkin-Huxley terms.

It has long been recognized that a Na pump is necessary to reverse the dissipative movement of Na involved in the generation of an action potential in nerve. If one could add to this requirement that of a Ca pump to reverse the dissipative movements of Ca, it might be possible to explore cardiac electrogenesis not by measuring the highly complex dissipative currents, but by analyzing the somewhat simpler restorative currents. The movement of K is largely ignored, since a combination of anomalous rectification and a membrane potential close to E_K (except in nodes) combine to make this sort of dissipative ion movement small compared with those of Na and Ca.

Much of the information about restorative ion movements (i.e., transport) comes from recognizing that two separate mechanisms are capable of producing a Na efflux. The first is the conventional Na/K pump; the second is the Na/Ca transport mechanism. Since the latter mechanism is usually thought of as pumping out Ca in exchange for the entry of Na, it is helpful to describe the Na efflux produced by this transport as a "running backward." While the Na/K pump has been caused to "reverse" and produce ATP while dissipating Na and K gradients, the requirements for this mode of operation are not likely to be met under physiological conditions. Thus Na/K exchange may be viewed as always running forward, while changes in membrane potential and $[Ca]_i$ and $[Na]_i$ are capable of reversing Na/Ca exchange.

The purpose of this volume is to emphasize the role of Na/Ca exchange in accomplishing the following during each heartbeat: (a) stopping the outward transport of Ca and initiating an inward Ca transport during the plateau, and (b) reversing the direction of carrier-mediated Ca transport during the diastolic interval so that not only the Ca gain from (a) above but also the Ca gain from I_{Ca} moving through slow inward channels is reversed. A role for Na/Ca exchange is present in the initiation of both contraction and relaxation. For Na/Ca exchange to behave in this way, the system must be electrogenic (i.e., transport more than two Na per Ca); the evidence for such an electrogenic mode of operation is listed below and is discussed in detail in subsequent chapters:

1. A $[Ca]_i$ of 20 to 50 nM that is appropriate to a relaxed muscle fiber can be produced only by a coupling ratio of four Na to one Ca (1).
2. Ca efflux in squid axons (2) varies as $\exp(-E_m F/RT)$, with F representing Faraday, R a gas constant, and T temperature. Ca influx that is coupled to Na exit varies as $\exp(E_m F/RT)$. This is appropriate voltage sensitivity of a carrier moving two net charges per cycle.
3. Measurements analogous to the preceding point made by depolarizing (3) cardiac fibers with KCl are not satisfactory. Such treatment changes the ^{45}Ca specific activity by the entry of unlabeled Ca and also makes changes in $[Ca]_i$ by the release of Ca from the sarcoplasmic reticulum (SR).
4. Small changes in $[Na]_i$, such as those produced by clinical doses of ouabain, yield large changes in Ca entry (4). Since at constant E_m the entry of Ca is given by $[Ca]_0([Na]_i)^r$, where

r is the coupling ratio Na/Ca, a value for r in excess of 2 is indicated by these measurements (1).

5. Tension continues to increase in cardiac fibers clamped at or beyond E_{Ca}. The evidence is convincing that the Ca producing contraction comes from outside the fiber (5); only a nonelectroneutral Na/Ca exchange can explain this effect.
6. Membrane vesicles isolated from cardiac myocardium and studied *in vitro* show an electrogenic Na/Ca exchange (6).

An important role for Na/Ca exchange in influencing contraction was suggested by the pioneering work of Repke (7) and at the same time by Langer (8); this was 4 years before the classic paper of Reuter and Seitz (9) that suggested a Na/Ca countertransport. Problems exist in accepting the idea that Na/Ca exchange might move Ca both into and out of cardiac fiber. First, it is difficult to make the isotopic flux measurements necessary to prove the hypothesis; these require squid giant axons. Second, skeletal muscle physiology, with its dominant SR Ca release, has unduly influenced thinking about cardiac fibers.

A further difficulty in initiating transport studies may have been that voltage clamp methods were applied to cardiac fibers long before an appropriate analysis of the errors that might be involved. Thus a substantial literature concerning artifacts was built up and only recently have there been realistic efforts to analyze ionic currents in cardiac fibers.

There is at present great interest in the extent to which Na/K pumping can affect the parameters of cardiac contractility; only a short while ago electrical measurements and ion transport studies were separate fields, with little interconnection. It is now known that not only Na/K transport with its contribution to membrane potential and to $[Na]_i$ but also Na/Ca exchange may be intimately involved with excitation-contraction coupling. This presages a new era of research that will address not only dissipative ion movements but also restorative processes.

The quantitative aspects of Na/Ca and Na/K exchanges cannot be discussed without assumptions about actual values for $[Na]_i$ and $[Ca]_i$. First, we assume that $[Na]_i$ is 10 mM in resting, that is, noncontracting, myocardium, and may be 20 mM when the heart is beating and hence subject to a substantial Na load. These assumptions do not differ from Na electrode measurements of an activity for Na^+ of 7 mM and an activity coefficient of 0.75, which would make $[Na]_i$ = 9.3 mM in resting fibers. Substantial reservations about ion-specific electrode measurements are discussed in this volume. The rarefied thermodynamic concept of ion activity in cardiac fibers is perhaps best dismissed considering the rather primitive state of the art of its measurement. Recently, electrode measurements of $[Ca]_i$ have begun to appear. The greater difficulties of an experimental sort that exist here suggest the need for caution about embracing a particular value. Among these difficulties are the following: (a) calibration is with CaEGTA/EGTA buffer, yet realistic calibrating solutions are seldom used; the myoplasm has a high ionic strength since the principal anions are ATP^{4-} and CrP^{3-} and the dissociation of CaEGTA is sensitive to ionic strength; (b) the penetration of a myocardial fiber with a large electrode almost surely disrupts much SR with a consequent change in Ca buffering; (c) the leak induced by a large electrode in a myocardial cell must be expected to depolarize it and hence to lead to a larger steady state $[Ca]_i$ than normal; it also leads to a Na leak that is quantitatively much larger than any Ca leak, since Na_o/Ca_o is of the order of 75. Such a Na leak must be expected to perturb $[Ca]_i$.

The values used are for $[Ca]_i$ determined in the squid giant axon where none of these problems with electrodes can arise, and where the dissociation constant of CaEGTA has been determined at the precise ionic strength existing in axoplasm. This involves values of 30 nM for the $[Ca]_i$ in resting fibers and higher values as $[Na]_i$ is increased. None of the arguments about Na/Ca exchange critically depends on exact values of $[Na]_i$ and $[Ca]_i$ but only the potential where reversal of Na/Ca exchange takes place.

The following notation is used: [ ] denotes free ion concentration, where the subscripts i and o refer to inside and outside the membrane, respectively; but Na_i and Na_o, for example, are employed to represent internal and external ionic species where no reference to a concentration is intended. Currents I_{Na} and I_{Ca} are the actual net fluxes of the particular ions, whereas I_{si} is the mixed current through a particular channel. A new current, I_C, is introduced for the carrier current of Na/Ca exchange. Note that in the balance of the text the ionic charges belonging to the different species are dropped but should be understood.

The presentation is organized first by a consideration of carrier-mediated transport in general, followed by a treatment of Na/K transport, Na/Ca transport, and the interactions among these systems. Finally, use is made of these considerations of ion transport to explain how the cardiac action potential is able to control the production of both tension and relaxation.

REFERENCES

1. Mullins, L. J. (1976): *Fed. Proc.,* 35:2583–2588.
2. Mullins, L. J., and Brinley, F. J., Jr. (1975): *J. Gen Physiol.,* 65:135.
3. Jundt, H., Porzig, H., Reuter, H., and Stucki, J. W. (1975): *J. Physiol. (Lond.),* 246:229–253.
4. Bennett, M. R., and Lavidis, N. A. (1979): *J. Gen. Physiol.,* 74:429–456.
5. Anderson, T. W., Hirsch, C., and Kavaler, F. (1977): *Circ. Res.,* 41:472–480.
6. Pitts, B. J. R. (1979): *J. Biol. Chem.,* 254:6232–6235.
7. Repke, K. (1964): *Klin. Wochenschr.,* 41:157–165.
8. Langer, G. A. (1964): *Circ. Res.,* 15:393–405.
9. Reuter, H., and Seitz, N. (1968): *J. Physiol.,* 195:451–470.

Chapter 2

Carrier-Mediated Transport

DIFFUSION VERSUS CARRIERS

Channels

The large currents measured across the excitable membrane by electrophysiologists have required the development of a structure that could accomodate such ion traffic. Phospholipid bilayers are not a promising medium for ion movement since their specific resistivity is high; the incorporation into a bilayer of a protein whose core is predominantly aqueous while its periphery remained lipid-like makes possible the channel concept. Channels can be gated, meaning that stimuli can open or close the channel to ion traffic, or they can be continuously open structures. Diffusion in the structures is approximately that given by Fick's law.

Carriers

An older concept in membrane studies is that of a carrier; it is an organic molecule that is highly lipid soluble and hence remains in the membrane phase. Again, the idea is to provide a way around the diffusion barrier imposed by the phospholipid bilayer. Originally, carriers were ferries that crossed back and forth carrying substances from one side of the membrane to the other. With the isolation proteins with sizes of 250,000 daltons, this arrangement is less plausible. With the further requirement that carriers have binding sites on both sides of the membrane that must be filled at the same time, there is more of an inclination to regard carriers as continuous from one side of the membrane to the other. The difference between channels and carriers may be only that we assign relatively slow movements and those with high specificity to carriers, while structures allowing more rapid and less specific solute movements are called channels. Diffusion by carrier is easily saturable, a property not associated with channels.

FACILITATED DIFFUSION

The movement of substances across the cell membrane is via the intervention of a carrier or substance that is confined to the membrane phase and combines with the carried species of ion or molecule at one membrane surface and releases the species at the other membrane surface. For the reactions shown below, which are those involved in transport, X is the carrier; S is the substrate to be carried; as before, the subscript i stands for inside; and o, outside.

$$S_o + X_o \rightleftarrows SX_o \qquad [1]$$

$$SX_o \rightleftarrows SX_i \qquad [2]$$

$$SX_i \rightleftarrows S_i + X_i \qquad [3]$$

$$X_i \rightleftarrows X_o \qquad [4]$$

Reaction 1 is the formation of the diffusible carrier complex; 2 is the translocation of the complex, 3 the dissociation of the complex, and 4 the return of the unloaded carrier. If reaction 4 is made impossible, the reaction is called exchange diffusion since the fluxes of S in both directions are required to be equal. Reaction 1 can be defined by

$$K = \frac{([S]_o)\,([X]_o)}{[SX]_o} .$$

Since it is usually considered that the translocation reaction (2) is the rate-limiting step, it is the concentration of SX_o that will determine the flux, or

$$\text{Flux} = \frac{([S]_o)\,([X]_o)}{K}\,.$$

For low concentrations of S_o, $[X]_o$ remains approximately constant, so that flux increases linearly with $[S]_o$; for large values of $[S_o]$, saturation of the system will occur. The usual arrangement is: $[X]_o + [SX]_o = 1$; thus $K/[S]_o = (1-[SX]_o)$, where the term on the left is a reduced concentration. When $K = [S]_o$, then $[SX]_o = 0.5$, or the flux is half-saturated. Note that this mechanism will catalyze the dissipation of the gradient in S across the membrane, and a net flux of S will cease when $[S]_o = [S]_i$.

COTRANSPORT AND COUNTERTRANSPORT

In sharp contrast to the Na pump with its requirement for a substrate, ATP, there also exist in a variety of cells transport mechanisms that move sugars, amino acids, ions, and other substances in an uphill direction using as an energy source the Na electrochemical gradient. A single, ATP-consuming transport system, the Na/K pump, establishes the Na electrochemical gradient. Other modes of transport then draw on this secondary source of free energy for their requirements.

Systems are known in which Na^+, a sugar (S), and a carrier (X) combine in a ternary complex [Na-X-S]. This diffuses across the membrane and dissociates. The result is an increase in $[Na]_i$, which is rapidly reversed by the Na pump, and an increase in the concentation of S_i. This is cotransport and is a common feature of secretory epithelia, where substrates must be conserved. In some cases, the driving ion (Na^+) and the substance or ion to be transported are on opposite sides of the membrane. Thus when Na^+ enters, an ion from inside the cell is moved outward. In this case, the process is called countertransport. An example might be the efflux of Mg coupled to the entry of Na (1).

How much energy is available in the Na electrochemical gradient? The chemical part of the gradient is $\mu_C = -RT \ln([Na]_o/[Na]_i)$, while the fact that there are energetic consequences of the diffusion of a charged particle means that the electrical part of the gradient is $\mu_E = zFE$ (z represents ionic valence). Thus the total electrochemical gradient for Na is $\mu_C + \mu_E$, or $\mu_{EC} = -RT \ln([\text{Na}]_o/[\text{Na}]_i) + zFE$. For many purposes, it is convenient to express energies in terms of potential because membrane potential often is measured. Therefore,

$$\frac{\mu_{EC}}{zF} = -\frac{RT}{zF} \ln \frac{[\text{Na}]_o}{[\text{Na}]_i} + E$$

where E is the potential difference (or membrane potential) between the phases involved in the measurement. Since the first term in the above equation is defined as E_{Na}, the potential at which Na^+ are in equilibrium, the expression is equivalent to defining the energy of the Na gradient as $zF(E - E_{\text{Na}})$. This treatment of the energy in the Na gradient is discussed in detail because it is important to remember that the gradient consists of two terms; and an increase in either can increase the size of the gradient. With respect to the kinetics by which Na^+ move, these may not necessarily be equivalent; for thermodynamics, however, they are equal sources of energy.

How much energy does the transport to be driven require? If we consider a sugar transport, then the energy required is $-\text{RT} \ln([\text{S}]_i/[\text{S}]_o)$, where [S] is the sugar concentration. Thus at equilibrium and with no inefficiency in the transport coupling,

$$-RT \ln([\text{S}]_i/[\text{S}]_o) = -RT \ln([\text{Na}]_o/[\text{Na}]_i) + zFE$$

or

$$[\text{S}]_i/[\text{S}]_o = ([\text{Na}]_o/[\text{Na}]_i) \exp(-zFE/RT)$$

With the usual values of $[\text{Na}]_o/[\text{Na}]_i$ in cells of about 10, and with a membrane potential of -75 mV, $\exp(-zFE/RT)$ is about 20. Thus the accumulation ratio S_i/S_o is in the limit of about 200. We conclude, therefore, that the Na electrochemical gradient is capable of concentrating inside a cell a nonelectrolyte, such as sugar or amino acid, by a value of 200-fold that of its external concentration using the energy stored in the Na gradient produced previously by Na pumping.

Reactions for cotransport can be written thus:

$$Na_o + X_o \rightleftharpoons NaX_o \rightleftharpoons NaX_i \rightleftharpoons Na_i + X_i \quad [1]$$

$$S_o + X_o \rightleftharpoons SX_o \rightleftharpoons SX_i \rightleftharpoons S_i + X_i \quad [2]$$

$$NaX_o + S_o \rightleftharpoons NaX_oS \rightleftharpoons NaX_iS \rightleftharpoons Na_i + X_i + S_i \quad [3]$$

$$SX_o + Na_o \rightleftharpoons NaX_oS \rightleftharpoons NaX_iS \rightleftharpoons Na_i + X_i + S_i \quad [4]$$

$$X_T \rightleftharpoons X_o + X_i + NaX_o + NaX_i + NaX_oS + NaX_iS + SX_o + SX_i \quad [5]$$

The first two reactions allow a dissipation of the Na and S gradients, respectively. If they are not suppressed, the transport will be inefficient. Reactions 3 and 4 are equivalent, although the kinetic parameters of the system may greatly favor one reaction over another. Reaction 5 is a conservation of carrier statement and is important in analyzing transport; under certain conditions, much of the carrier can be tied up in slowly moving or nonmoving forms (reactions 1 and 2).

The form of the flux equations will be the product [Na] × [S] × [X], with the influx given by values for the ion, sugar, and carrier on the outside, and the opposite for efflux.

There is some evidence (2) that Na_o may exchange for H_i via a countertransport mechanism. The process is likely to be more complicated, but it is informative to consider how a countertransport might be effected by the Na electrochemical gradient. As mentioned above, the Na gradient can be written as $(E - E_{Na})zF$, while that for the energy required for the transport of H^+ can be written as $(E - E_H)zF$. Equilibrium will occur when $(E - E_H) = (E - E_{Na})$ or when $E_{Na} = E_H$. With a Na ratio of 10, $E_{Na} = +58$ mV; thus countertransport would in fact make $E_H = +58$ mV, while experimentally it is closer to zero. If H^+ were distributed passively, then $E_H = E_m = -75$ mV, or the cell interior would be 20-fold more concentrated in H^+ than would the external medium. Another possibility is that more than 1 H^+ exchanges for 1 Na^+, so that $n(E - E_H) = (E - E_{Na})$, where n is the number of H ions per Na ion on the exchange carrier. For purposes of present illustration, if $E = -75$ mV and $E_{Na} = +58$ mV, a solution to the above equation with $n = 2$ is $E_H = -8$, a result in substantial agreement with the experiment. Note that the operation of this carrier stoichiometry involves the generation of an electric current, since more charges move in one direction than in another per cycle of the transport mechanism. The efflux of H^+ in exchange for Na^+ means that more charges leave the cell than enter; thus the movement is hyperpolarizing.

Another countertransport that has been studied in the squid giant axon is that for Mg. It was found that Mg efflux is promoted by Na_o and inhibited by Na_i, while the converse is true for Mg influx. Measurements of $[Mg]_i$ suggest a value of about 2.5 mM for free, ionized Mg, and a value for $[Mg]_o$ of about 25 mM. These values are uncertain to the extent that both Mg_o and Mg_i are complexed with substances such as SO_4 and ATP, respectively. These values mean that $(E - E_{Mg})z = (-75 - 29)2 = (E - E_{Na})r$, where r is the ratio Na/Mg. Clearly, equality in this instance demands a value of 2 or more for r. A value of 2 would allow an electroneutral exchange of Mg for Na. In the case of Mg efflux: $[Mg]_i \times ([Na]_o)^2 \times [X]$. The square in the Na term enters because apparently 2 Na interact per 1 Mg transported; X can be either X_o or X_i, depending on how the model is structured with respect to the binding first of Na or of Mg. Thus it is possible to have values of r of less than unity (the illustration for H^+ where $r = 0.5$; $n = 2$) and an electrically active transport. It is also possible to have an electroneutral transport as outlined above. Note, however, that all the sugar and amino acid transport systems operating on the Na gradient are electrogenic, since their operation involves the uncompensated movement of charge from Na^+.

The general importance of Ca in cell function has been widely appreciated; there have been many efforts to understand how [Ca] is regulated in cells. Progress in understanding Ca regulation has been hindered severely by our inability to measure $[Ca]_i$, the ionized

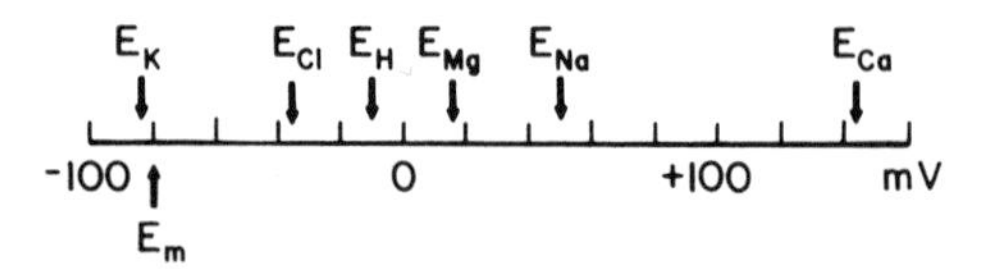

FIG. 2.1. Equilibrium potential for ions present in cardiac fibers. Based on the analytic data given in Table 5.1.

free [Ca] inside cells. As long as one could assume a free Ca inside cells of the order of 20 μM, a countertransport of 2 Na/Ca could be assumed, since this would produce an equilibrium concentration ratio of Ca across the membrane of 100 if the Na ratio were 10. This effect arises because at equilibrium, $\mu_{EC}(\mathrm{Ca}) = \mu_{EC}(\mathrm{Na}) \times r$, where r is the coupling Na/Ca. Hence

$$-(RT \ln([\mathrm{Ca}]_o/[\mathrm{Ca}]_i)) + zFE = -(RT \ln([\mathrm{Na}]_o/[\mathrm{Na}]_i)) \times 2 + 2zEF$$

The electrical terms are equal and cancel for electroneutral exchange, so that

$$[\mathrm{Ca}]_o/[\mathrm{Ca}]_i = ([\mathrm{Na}]_o/[\mathrm{Na}]_i)^2$$

The extent to which various ions are in need of energy for their transport is shown by the diagram in Fig. 2.1. With a resting membrane potential as indicated (and near E_K), the ion requiring the greatest expenditure of free energy for its transport is Ca. The actual energy requirement for this ion is twice the distance between the membrane potential and E_{Ca}, since the particle being transported is double charged. For Mg, transport requirements are more modest, because its concentration inside cells is of the order of millimolar rather than tens of nanomolar. H^+ has already been referred to, and Cl^- has modest requirements for its transport.

REFERENCES

1. Mullins, L. J., Brinley, F. J. Jr., Spangler, S. G., and Abercrombie, R. F. (1977): *J. Gen. Physiol.*, 69:389.
2. Thomas, R. C. (1976): *Nature*, 262:54–55.

Chapter 3

The Na/K Pump

HISTORY

At least in the cardiac fiber, it is clear that in addition to providing the usual membrane hyperpolarization and reversal of Na leaks, the Na/K pump plays an exceedingly important regulatory role in determining the strength of contraction. It is in effect coupled to the Na/Ca exchange system by virtue of each pump sharing a common ion, Na^+.

The original suggestion of Na pumping was made by Dean (1), who considered that since isotopic measurements had shown that Na entered cells, an energy-consuming mechanism was necessary to reverse this movement. This original proposal was for an electrogenic (i.e., charge-producing) pump. Experiments quickly showed that Na efflux as measured with isotopes was highly sensitive to the presence or absence of K in the external solution. It was logical to suppose that the mechanism being studied exchanged Na for K or was electroneutral. In the case of most animal cells, and especially in muscle, membrane potentials were so close to E_K that an inward K pumping was hardly necessary; but in the case of red blood cells (RBC), then undergoing intensive study, the virtual absence of a membrane potential made an inward K flux by pumping a necessity.

The next development was a more quantitative measurement of K and Na fluxes and the discovery by Schatzman (2) that cardiac glycosides were specific inhibitors of the Na/K pump. These developments showed that there were always more Na extruded than K taken up; thus the mechanism was again electrogenic. For RBC, a ratio of 3:2 was measured under physiological conditions; for other tissues, values of 3:1 could also be measured.

A final phase of Na/K studies was the demonstration that the pump could run in a number of different modes whereby it would exchange Na/Na, Na/K, K/K, Na/ — that is, an uncoupled Na efflux—and could also be reversed so that it was K/Na in its exchange (3).

SUBSTRATE

The net movement of Na from the interior of most animal cells to the external solution where Na is approximately 10-fold more concentrated requires the performance of work. As noted in Chapter 2, this work can be done by using either an existing solute gradient as an energy source or substrates formed in the cell by metabolism. In the case of the Na pump, experiments with RBC and resealing techniques made it clear that ATP was in fact the essential substrate for (a) the operation of the Na pump and (b) hydrolysis by an enzyme that was activated by Na and K and inhibited by cardiac glycosides.

In the case of squid giant axons, it was possible to remove by internal dialysis all the substrates in the fiber and to then introduce a variety of single substrates. Thus it was shown that only ATP, and to a lesser extent deoxy-ATP, was capable of energizing the Na efflux (4). Other experiments have shown that 3 Na^+ are transported per ATP hydrolyzed and that the movement of Na from a fiber is substantially independent of membrane potential.

ACTIVATION BY IONS

The Na pump will not extrude any ion other than Na when this is presented to the mechanism at the inside of the cell. On the other hand, a variety of external cautions have a stimulating effect on the rate of Na extrusion. Indeed, K, Rb, and NH_4 all have about the same action in promoting Na extrusion, whereas Li and Cs have smaller stimulating roles. External Na appears to compete with K for activation of the Na pump. Thus reductions in $[Na]_o$ lead to increases in Na pumping at a given [K]. A change in $[K]_o$ is sometimes used experimentally to either promote or reduce the rate of Na pump operation. This ion has an effect that is about half-maximal when $[K]_o$ = 1 mM (given a normal $[Na]_i$). The relationship between $[K]_o$ and Na efflux is an S-shaped curve with saturation around 2 mM, as shown in Fig. 3.1. Since Na-free conditions are often imposed on cardiac fibers under experimental study, it should be recognized that any membrane potential change could be the result of increased stimulation of the Na/K pump; removal of Na_o allows K_o to be more effective in pump stimulation. The Na/K pump requires Na_i, ATP_i (as the Mg complex), and K_o for its maximal activation; other factors, such as Na_o, can modulate the rate of pumping but are not essential.

CONCENTRATION DEPENDENCE ON Na_i

The extrusion of Na from cells is a necessity if both osmotic balance and a membrane potential are to be maintained. The existence of a particular value for Na influx also specifies a value for

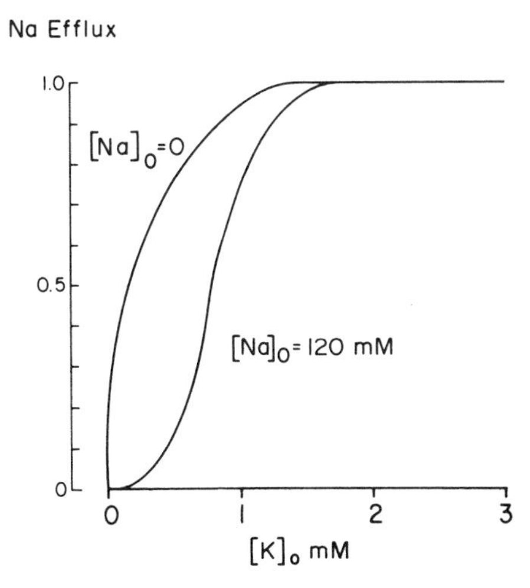

FIG. 3.1. Stimulation of Na efflux by K_o when $[Na]_o$ = 120 mM or 0. In the first case, the Na pump is half-saturated at $[K]_o$ = 1 mM, while in the absence of Na_o, half-saturation is at $[K]_o$ = 0.1 mM.

$[Na]_i$ at which the Na efflux generated by the Na pump equals that of Na influx. Transient values for Na pumping can deviate from this flux balance, but $[Na]_i$ will change until fluxes are balanced.

There is a curious difference between human RBC and both muscle and nerve in the way that Na efflux from these various cells responds to changes in $[Na]_i$. In the case of RBC, this relationship is an S-shaped one with a value for half-saturation of about 30 mM Na_i. In the case of muscle, Na efflux has had to be measured in Li Ringer's in order to avoid unknown changes in specific activity. Here the relationship is steep (greater than 3.5 power function of Na_i) over a limited low range of $[Na]_i$. By contrast, in nerve cells, Na efflux is linear with $[Na]_i$ over a wide concentration range. This puzzling behavior of the Na pump remained unexplained until it was found (5) that part of the Na efflux from squid axons depended on $[K]_o$ (the Na/K pump) and was sensitive to ouabain, and part depended on $[Ca]_o$ and was insensitive to ouabain. This $[Ca]_o$-dependent Na efflux did depend on $[Na]_i$ and could be identified with an exchange of 1 Ca entering the fiber, while 4 Na exited, i.e., a coupled Na/Ca movement. Measurements of Ca influx have also identified a coupled Ca entry and Na exit as a mode of operation. Ca influx in squid axons has been shown (6) to have a K_{Na} for activation of about 60 mM Na_i, whereas the K_{Na} for half-activation of Na efflux by the Na/K pump in RBC[1] is of the order of 30 mM. Both curves are sigmoid as might be expected, since Na pumping is usually considered to take 3 Na per translocation, while the Na/Ca mechanism may take 4.[2]

From the point of view of cellular homeostasis, $[Na]_i$ must be such that over a working cycle of the cardiac fiber, the time integral of the sum of Na influx plus efflux equals zero; that is, $[Na]_i$ is constant. In an ordinary excitable nerve fiber, the sources of Na entry are a baseline Na influx plus an additional influx from Na entering via Na channels with each action potential. That is, one

---

[1]We must assume the value found in mammalian RBC, because in other tissues Na/Ca is likely to contribute to Na efflux versus Na_i.

[2]How these curves interact will be explored in more detail in Fig. 6.1 and the surrounding discussion.

term is independent of bioelectric activity, and one depends on it directly. In cardiac muscle, both types of Na entry and a term proportional to net Ca entry should be expected, because Ca entry via I_{Ca} must be reversed by further Na entry. Strong evidence suggests that associated with each heartbeat, there is not only a net Na entry from the fast upstroke of the action potential, but also a Na entry associated with Ca removal; this Ca removal is necessary to reverse the measured Ca entry via slow inward current. We can further expect some basal (background or leakage) Ca entry equivalent to the Na entry mentioned above; this will also require a further Na entry in order to reverse its effect.

ELECTRIC CURRENT GENERATION

Historically, notions about the mechanisms of the membrane potential included the following: it was (a) an equilibrium K diffusion potential, (b) a diffusion potential shunted by currents generated by ions not at equilibrium (the Goldman-Hodgkin-Katz relation), and (c) a potential as in (b) enhanced by an electrogenic component generated by metabolism. More recently, it has been realized that carrier-mediated ion movements in general may add to or subtract from the membrane potential.

One of the factors that limits speculation on the nature of the membrane potential is the necessity for flux balance in the steady state. For example, if $E_m = E_K$ [corresponding to (a) above], then K fluxes are in balance, and the inward movement of Na by diffusion is balanced by Na pumping so that the movement of Na across the membrane does not contribute current. This pump is electrogenic, since stopping the pump will result in membrane depolarization. In nerve fibers, however, the membrane potential was a somewhat less than plausible value for E_K, and $[K]_o$ had an influence on the Na pump that could best be described as a coupled movement of Na and K driven by metabolism. To explain the divergence of E_m from E_K, the assumption was made that currents from ions such as Na were not zero [as in (b) above], but that in fact the Na pump was an electroneutral mechanism moving Na and

K in equal quantities per pump cycle. This meant that Na entering the fiber constituted a current that would depolarize the membrane to an extent such that the outward K current and inward Na current would sum to zero. The loss of K and gain of Na would then be balanced by the 1:1 coupled Na/K pump. This is equivalent to

$$I_K + I_{Na} = 0$$

$$I_K = P_K F([K_o] - [K_i] \exp(EF/RT))\, f(E)$$

where P is permeability and $f(E)$ is some function of membrane potential.

$$I_{Na} = P_{Na} F([Na_o] - [Na_i] \exp(EF/RT))\, f(E)$$

$$E = \frac{RT}{F} \ln \frac{P_K[K]_i + P_{Na}[Na]_i}{P_K[K]_o + P_{Na}[Na]_i}$$

This proved an adequate scheme until more precise measurements of Na pump fluxes showed that Na efflux was substantially larger than K influx. Furthermore, in Na-loaded frog muscle fibers, Kernan (7) was able to show that the membrane potential was transiently substantially higher than E_K when the muscles were allowed to unload Na. These new findings required a coupling ratio Na/K of greater than 1. In RBC, the ratio appeared quite constant at 3:2, whereas in squid axons, it was closer to 3:1 if the pump were defined as the ATP-dependent fluxes of Na and K. In addition, K influx dependent on ATP appeared to saturate, while Na efflux dependent on ATP did not; hence the coupling ratio could be very large if $[Na]_i$ were high. Again, in the steady state, pumping must operate such that the currents of Na and K and that of the pump sum to zero.

REFERENCES

1. Dean, R. B. (1941): *Biol. Symp.*, 3:331.
2. Schatzman, A. J. (1953): *Helv. Physiol. Pharmacol. Acta*, 11:346–354.
3. Glynn, I. M., and Karlish, S. J. D. (1975): *Annu. Rev. Physiol.*, 37:405–459.

4. Benninger, C., Einwachter, H. M., Haas, H. G., and Kern, R. (1976): *J. Physiol. (Lond.)*, 259:617–645
5. Baker, P. F., Blaustein, M. P., Hodgkin, A. L., and Steinhardt, R. A. (1969): *J. Physiol. (Lond.)*, 200:431.
6. DiPolo, R. (1979): *J. Gen. Physiol.*, 73:91–113.
7. Kernan, R. P. (1962): *Nature*, 193:986.

Chapter 4

Na/Ca Exchange

DESCRIPTION

It has long been known that increasing $[Ca]_o$ or decreasing $[Na]_o$ would increase tension development in cardiac muscle. Quantitative analysis of this effect was first made by Wilbrandt and Koller (1) and further developed by Lüttgau and Niedergerke (2). These authors were impressed by the apparently competitive nature of these two ions in crossing the membrane. It was natural to propose that a carrier, R, might combine either with 2 Na or 1 Ca. Hence the idea that $[Ca]_o/([Na]_o)^2$ would determine the fraction of the carrier in the Ca-carrying form.

Further measurements using isotopes showed that Ca efflux from cardiac tissue depended on $[Na]_o$. In 1968, Reuter and Seitz (3) introduced the idea that a Na/Ca countertransport might be responsible for Ca movement in cardiac fibers. Meanwhile, Blaustein and Hodgkin (4) and Baker et al. (5) were able to show in squid axons that Ca efflux depended on $[Na]_o$, and that there was a large $[Ca]_o$-dependent Na efflux whereby 3 to 5 (mean, 4) Na^+ emerged per Ca entering the fiber. These results made it possible to envisage a single exchange mechanism that would move Ca both inward (as suggested by Lüttgau and Niedergerke) or outward via a carrier-mediated process. Part or all of the energy required could come not from conventional substrates, such as ATP, but from the Na electrochemical gradient already formed in the cell as a result of an ATP-dependent Na pumping.

The squid giant axon proved to be ideally suited to making the various measurements necessary to work out the quantitative relationships among the six variables known to affect Ca flux balance

(Table 4.1). Such a simplified tabulation cannot be entirely correct since many of the variables interact with each other. Thus $[Na]_o$ increases Ca efflux if $[Ca]_i$ is high but may have no effect if $[Ca]_i$ is low; ATP increases Ca efflux if $[Na]_i$ is finite but has no effect on Ca efflux if $[Na]_i = 0$. These interactions among variables are discussed below. First, the thermodynamics of Ca transport are considered.

ENERGETICS

The original suggestion of Reuter and Seitz (3) was that the countertransport mechanism exchanged 1 Ca for 2 Na; i.e., it was an electroneutral carrier. This followed from earlier observations that Na and Ca appeared to compete for entry into cardiac fibers and that the ratio of $[Ca]_o/([Na]_o)^2$ was constant for contractile activation. If

$$-RT \ln \frac{[Ca]_o}{[Ca]_i} + zFE = r\left(-\mathrm{RT} \ln \frac{[Na]_o}{[Na]_i} + zFE\right),$$

then for $r = 2$,

$$\frac{[Ca]_o}{[Ca]_i} = \left(\frac{[Na]_o}{[Na]_i}\right)^2.$$

Since the Na ratio across the membrane is about 10, the Ca ratio would be 100, or $[Ca]_i$ would be 1/100th of 2 mM, or 20 μM. The threshold for contractile activation of muscle is at approximately

TABLE 4.1. *Effect of six variables on Ca flux balance*

	Ca	
Increase in	Efflux	Influx
$[Na]_o$	+	−
$[Na]_i$	−	+
$[Ca]_o$	+	+
$[Ca]_i$	+	+
E_m	+	−
$[ATP]_i$	+	+

0.1 μM Ca. Likely values for resting $[Ca]_i$ are in the range of 20 to 50 nM. Clearly, other coupling ratios of Na/Ca or other methods of Ca pumping are necessary to bring sarcoplasmic [Ca] to reasonable values. The equations below show values of $[Ca]_o/[Ca]_i$ that are possible if *r*, the coupling ratio of Na/Ca, equals 3 or 4.

$r = 3, E_m = -75$:

$$\frac{[Ca]_o}{[Ca]_i} = \left(\frac{[Na]_o}{[Na]_i}\right)^3 \exp\left(-\frac{EF}{RT}\right) = 2 \times 10^4$$

$r = 4, E_m = -75$:

$$\frac{[Ca]_o}{[Ca]_i} = \left(\frac{[Na]_o}{[Na]_i}\right)^4 \exp\left(-\frac{2EF}{RT}\right) = 4 \times 10^6$$

Thus for $[Ca]_o = 2 \times 10^{-3}$M, the equilibrium value for $[Ca]_i$ equals 100 nM ($r = 3$) or 0.5 nM ($r = 4$). However, the presence of Ca leaks into the fiber makes it unlikely that $[Ca]_i$ reaches its equilibrium value. Therefore, $r = 4$ is more likely than $r = 3$. Either value profoundly affects the way that Ca fluxes may be expected to change with membrane potential. The existence of electrogenic Ca pumping means that a new look must be taken at Ca movement via the Na/Ca exchange system.

The treatment thus far is thermodynamic; regardless of the mechanism by which Ca is pumped out, the work required is the same: $-RT \ln([Ca]_o/[Ca]_i) + zFE$. Thus Na/Ca exchange is not a way of "saving" energy (ATP is used to pump Na rather than Ca), nor does it matter whether a carrier has a permanent charge, since the pathway by which work is done is irrelevant to thermodynamics.[1] These points are mentioned because of suggestions that a charged carrier (which can affect kinetics) somehow circumvents the work of Na/Ca exchange.

---

[1]One also needs to know the power output of Na/Ca exchange at physiologically relevant values for $[Ca]_i$ and membrane potential, because leaks of Ca inward must be overcome by pumping and the values we have given are for systems without ion leaks.

The expression for $r = 4$ can be rearranged by taking logarithms of both sides

$$2\,E_{Na} - E_{Ca} - E = 0$$

as a condition of equilibrium. Since equilibrium is also defined by the presence of a flux ratio of unity, a potential exists where Ca fluxes are in balance: in effect, a reversal potential for the Na/Ca carrier, which we may call E_R

$$E_R = 2\,E_{Na} - E_{Ca}$$

Since the carrier moves unequal quantities of charge per cycle (4 Na versus 1 Ca or 2 net +), a deviation of the membrane potential from E_R means that a current will be generated by the Na/Ca carrier. The exact current-voltage relationship of the carrier depends on model considerations, but the thermodynamic equation shows that changing E_{Na} or changing membrane potential are equivalent operations in setting the equilibrium value of E_{Ca}.

A second important conclusion is that since E_{Na} in contracting myocardium is likely to be of the order of +50 mV, while E_{Ca} during diastole is +140 mV, $(2E_{Na} - E_{Ca}) = -40$ mV, a potential about midway between the resting potential and the plateau. A consequence of this arrangement is that during diastole Ca is pumped out of the fiber; during the plateau, not only is outward pumping of Ca stopped, but an influx of Ca occurs by the extrusion of Na_i in exchange for Ca_o. This makes part of the Ca entry during depolarization sensitive to $[Na]_i$, since the presence of Na_i is essential for exchange with entering Ca.

Alternate views of how a low $[Ca]_i$ is maintained are as follows: (a) ATP affects the affinity of the Na/Ca system, such that it can move Ca outward at physiological levels of Ca_i even though only 2 Na exchange per Ca moved (6); or (b) there is a parallel ATP-driven pump in addition to Na/Ca exchange (7).

Two contrasting possibilities are diagrammed in Fig. 4.1. In Fig. 4.1A, an arrangement is proposed whereby the leaks of Na and Ca via channels shown at the bottom are compensated with the Na being pumped out via an ATP-dependent pump and the Ca pumped out in exchange for external Na. In Fig. 4.1B, a Ca pump is driven by ATP, and the Na/Ca exchange mechanism is driven backward, thus extruding internal Na in exchange for external Ca, a situation that may occur when $[Na]_i$ is high.

The RBC is an exception to most animal cells in that it has a clearly demonstrable ATP-driven Ca pump. This may be necessary because the RBC has virtually no membrane potential, whereas most cells have −60 to −90 mV. Such an arrangement means that $E - E_{Na}$ for a RBC is −60 mV instead of a value twice as large. A special case is the dog RBC, in which there is no measureable ouabain-sensitive Na efflux, yet the cell manages to produce a $[Ca]_o$-dependent Na efflux. This suggests that Na is pumped out at the expense of Ca entry, and that Ca is pumped out by an ATP-driven pump (Fig. 4.1B).

To decide whether ATP or the Na gradient in nerve and muscle drive Ca, it must be determined whether a low $[Ca]_i$ can be main-

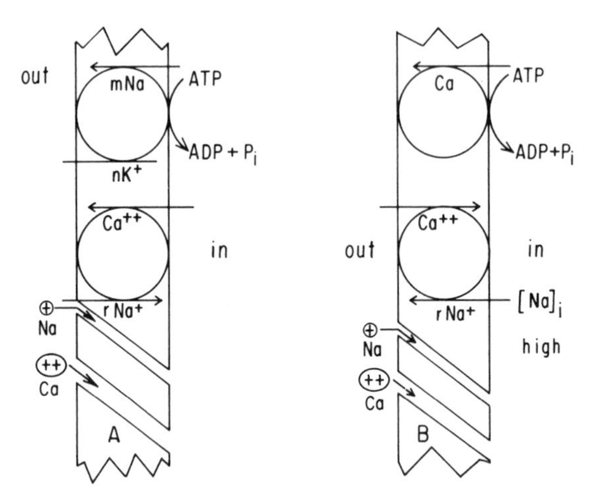

FIG. 4.1. Possible arrangement for reversing the dissipative movements of Na and Ca. **Left:** Na/K pump driven by ATP and Na/Ca exchange. **Right:** Hypothetical ATP-driven Ca pump to reverse the Ca leak and a Na/Ca exchange mechanism running backward and extruding Na in exchange for entering Ca. The latter arrangement may exist under conditions where the Na/K pump has been inhibited.

tained in the absence of ATP, provided that the Na gradient in both its electrical and chemical components is maintained. This eliminates direct experiments on cardiac muscle, since attempts to reduce ATP to zero inevitably allow a dissipation of the Na gradient. In squid giant axons, however, Na movement has a long time constant because of the large size of the fiber. Hence it is feasible to poison the ATP-generating machinery and still maintain E_m and Na_i at virtually normal values. Experiments measuring $[Ca]_i$ with aequorin (8) and total analytic Ca by atomic absorption (9) show that $[Ca]_i$ and analytic Ca are maintained at normal values in the absence of ATP; also, if axons are stimulated in high-Ca sea water or immersed in choline sea water (procedures that greatly increase both $[Ca]_i$ and total Ca), the axons will lose Ca when allowed to recover in 3 mM Ca sea water, even though ATP_i has been reduced to approximately 20 μM by poisons such as FCCP + IAA. If, on the other hand, a normal ATP is maintained in the fiber but the Na chemical gradient is abolished by making $[Na]_o$ = 40 mM = $[Na]_i$, then the axon does not lose its Ca load. If $[Na]_o = [Na]_i$ and the membrane potential is abolished with elevated $[K]_o$, then axons actually gain Ca. Evidently, any ATP-driven pump is unable to overcome the Ca leaks in the axon. The restoration of a normal Na electrochemical gradient can cause Ca loss such that both $[Ca]_i$ and analytic Ca approach control values with a time constant of 30 min.

If ATP were a factor in controlling the steady state $[Ca]_i$ of the axon, two arrangements would be possible: (a) a series arrangement, whereby ATP assisted Na/Ca exchange, and (b) a parallel arrangement of an ATP pump independent of the Na/Ca system. The first entails the difficulty that Ca flux measurements in squid axons show that ATP has no effect on Ca efflux if $[Na]_i = 0$ (10,11). Effects of ATP on Ca efflux are prominent only if $[Na]_i$ is high and $[Ca]_i$ is low. A study of Ca influx in squid axons (12) further showed a substantial stimulation of Ca influx by ATP as well as by $[Na]_i$. While ATP substantially increases Ca fluxes, it is incapable of producing a net flux of Ca. It remains to be experimentally demonstrated that in any tissue except the RBC, ATP produces a net Ca flux.

The parallel arrangement of a Na/Ca exchange and an ATP-driven pump suffers from another sort of difficulty. Na/Ca exchange is an ion movement whose direction is dictated by the direction of the Na electrochemical gradient, relative to that for Ca. If $[Ca]_i$ is substantially lower than the equilibrium value for a 2 Na/Ca exchange (i.e., 1/100th of $[Ca]_o$), then Ca_o will move inward in exchange for Na_i until $[Ca]_i$ approaches the value set by equilibrium (20 μM). The assumed parallel ATP-driven pump must then pump Ca strongly to overcome not only the Ca leaks but Na/Ca exchange as well, which is also providing a Ca gain. The parallel system described by DiPolo and Beauge (7) has a maximal flux in squid axons of ~ 0.1 pmoles/cm^2 compared with Na/Ca exchange fluxes of ~ 3 pmoles/cm^2. Such a pump cannot overcome large fluxes that are necessarily generated in cardiac muscle; yet such a transport system may engage in Ca pumping when cardiac muscle is Na_i, Na_o free. Under these circumstances, cardiac fibers can continue beating and developing action potentials (in solutions of elevated $[Ca]_o$ or upon treatment with norepinephrine), even though Ca extrusion and entry via Na/Ca exchange is not possible.

EFFECT OF ATP ON Ca FLUXES

As indicated earlier, ATP may drive a Ca pump with a small Ca efflux. In addition, experiments with squid axons make it clear that ATP has significant effects on fluxes of Ca generated by Na/Ca exchange. These effects are summarized in Fig. 4.2 (13) and show that in Fig. 4.2A there is little effect of ATP on Ca efflux when $[Ca]_i$ is in the range of 0.1 to 1.0 μM, while at resting levels of $[Ca]_i$, there may be a 10- to 20-fold increase in Ca efflux in the presence of ATP. The concentration of ATP that produces these increases in Ca efflux shifts toward higher values as $[Ca]_i$ rises (Fig. 4.2B). The effect of $[Na]_i$ on Ca efflux in the presence and absence of ATP and the Ca efflux plotted against $[Na]_o$ is shown in Fig. 4.2C. With $[Na]_i = 0$, the efflux of Ca is ATP independent; if $[Na]_i = 100$ mM, there is a large effect of ATP on Ca efflux. The details of this $[Na]_i$ effect are shown in Fig. 4.2D.

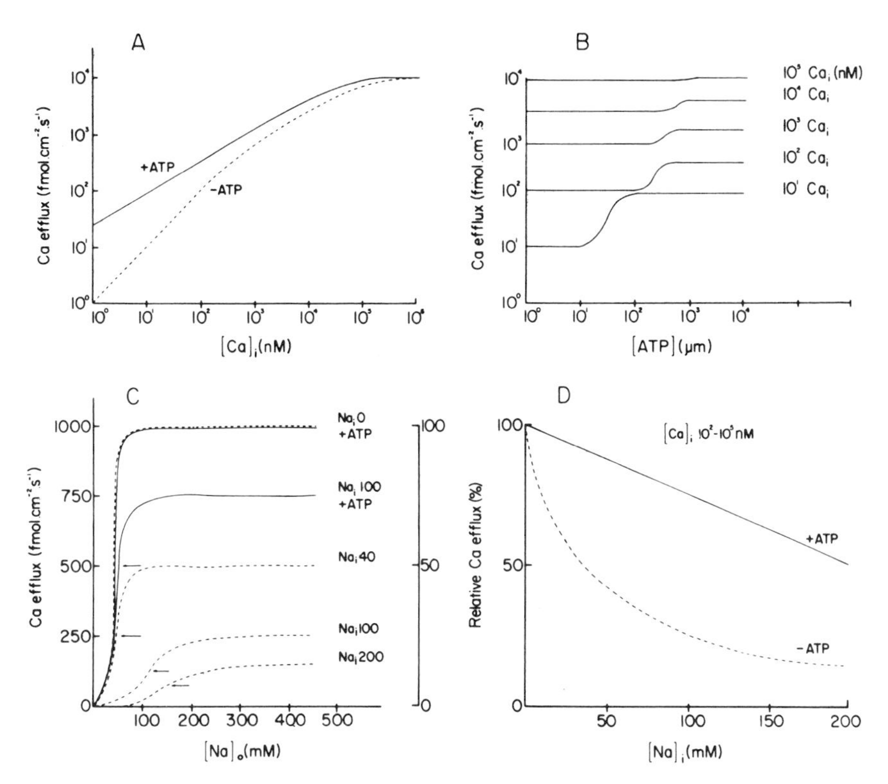

FIG. 4.2. Ca efflux in squid axons shown as functions of $[Ca]_i$ **(A)**, $[ATP]_i$ **(B)**, $[Na]_o$ **(C)**, and $[Na]_i$ **(D)**. In each case, the relevant curve shows how ATP affects the Ca flux. (From Requena and Mullins, ref. 13.)

The results outlined in Fig. 4.2 are compatible with the idea that ATP has an effect on the affinity of the Na/Ca carrier for Na. Other explanations are not ruled out, and the main conclusion is that since ATP has been shown to activate both efflux and influx of Ca, no effect is expected of this substance on net flux at equilibrium. On the other hand, since ATP increases unidirectional fluxes, if non-equilibrium conditions obtain, a large flux is expected in the presence of ATP and a smaller one in its absence.

The large degree of Ca and Na regulation possible in a normal cardiac fiber requires that the coupling ratio of Na to Ca be sufficiently high so that the Na electrochemical gradient alone is adequate to maintain measured values for $[Ca]_i$. The work to be done is $zF(E - E_{Ca})$, or for a resting membrane potential, E, of -80 mV and E_{Ca} at $+140$, this is $-440F$ mV. The energy available in the Na gradient is $zF(E - E_{Na})$, or $-80 - (+40)$ or $-120F$ mV; the coupling ratio of Na to Ca must be 440/120 or 3.67. This calculation assumes that the resting membrane potential is that present in the membrane most of the time. In reality, a cardiac fiber is depolarized to about zero membrane potential for one-third of the cardiac cycle. A proper estimate of membrane potential is to say that it is $(0.33 \times 0 + 0.67 \times -80)$ or -54 mV as an effective mean value. This calculation is an approximation, since we do not know the relationship between membrane depolarization and Na/Ca transport. With this mean value, the work to be done is $zF(-54 - 140)$ or $-388F$, while the energy available to drive the transport is $zF(-54 - 40)$ or -94 mV; this is a ratio of 4:1. In both instances, the calculation suggests that a value of 4 is the nearest integral minimal value for the coupling of Na to Ca.

Since the literature supports an electroneutral Na/Ca system, it is important to mention how this notion came about and what experimental support there is for electroneutral movement. It has been known since the time of Sidney Ringer that frog heart would contract in Na-free solutions. This finding was analyzed in some detail by Wilbrandt and Koller (1) as a carrier type process, whereby Na prevented Ca entry into the cardiac cell. Ten years later, Lüttgau and Niedergerke (2) were able to show that many of the data on contraction of frog atrial fibers could be reduced to a model where either 1 Ca or 2 Na competed for a site on the outside of the membrane. If the successful competitor were Ca, then this ion would enter and add to the pool of Ca_i available for and capable of causing muscle contraction. If, on the other hand, Na were the successful competitor, no contraction could result. Since the carrier had to have two sites, these could be occupied either by 2 Na or by 1 Ca. Over a certain range of concentration, a relationship was anticipated in which $[Ca]_o/[Na]_o^2$ should be a constant. The literature involving such measurements is enormous, but a critical review shows that the most uncritical measurements are encompassed in support of this rule. In studying responses at very low $[Ca]_o$ (14), it is clear that the rule fails and that $[Ca]_o$ itself is the variable that influences mechanical response. Similarly, if this ratio is examined in fibers depolarized to around E_{Ca}, this relationship again fails; a 4 to 6 power of $[Na]_o$ is necessary to encompass the experimental results (15). It is clear that experiments (16) specifically designed to test the $[Ca]_o/[Na]_o^2$ rule did not consider that if one elevates $[K]_o$ to depolarize the cell membrane, one (a) increases the release of Ca from the SR and hence increases $[Ca]_i$; (b) increases cold Ca entry via slow inward channels, hence decreasing the specific activity of ^{45}Ca inside the fiber; and (c) may affect the operation of the Na/Ca exchange system. It is difficult to conclude from such observations that Ca/Na stoichiometry is electroneutral. The evidence suggesting a nonelectroneutral exchange of Na for Ca is based on (a) the requirements of energetics for the process, (b) the fact that depolarization increases the mechanical response of cardiac fibers even though one approaches E_{Ca}, and (c) the fact that Ca entry and exit can be affected by a single entity; the Na/Ca exchange carrier controlled by three variables, E_{Na}, E_{Ca}, E_m, is a highly unifying concept.

CARRIER FLUX RATIO

The assertion is often made that a coupling ratio of 3 Na/Ca is adequate to maintain $[Ca]_i$ below the threshold for mechanical activation. This is probably so if $[Ca]_i$ can reach equilibrium in the time available between beats. There is always, however, some continuous nonspecific Ca leakage into fibers; hence, a steady state is achieved when (leak influx + carrier influx) = (carrier efflux), a condition that is not equilibrium for the Na/Ca system. Moreover, cardiac fibers often exhibit behavior most easily explained by assuming that the amount of stored Ca has been greatly changed. The easiest way to explain changes in stored Ca is to assume that stored Ca is related to $[Ca]_i$ and that it is possible to lower $[Ca]_i$ (and hence stored Ca) greatly below the level corresponding to the mechanical threshold. This matter will be discussed in detail in Chapter 7; here we will examine the consequences of (a) carrier equilibrium, (b) carrier disequilibrium in the presence of a leak, and (c) the effects that ATP may have if it promotes increased fluxes without a change in flux ratio.

There is some continuous inward Ca flux via leakage or background currents. A discontinuous Ca influx with each heart beat represents a further dissipative Ca movement. The energetic considerations of the previous section can be summarized (for the squid axon where the actual measurements have been made) as follows.

Ideal Na/Ca Countertransport

In this system, Ca can move across the cell membrane only via the countertransport carrier. At equilibrium, the free energy change is 0, and the Na and Ca gradients are related by the expression

$$rzF(E_{\mathrm{Na}} - E_m) = zF(E_{\mathrm{Ca}} - E_m)$$

The flux ratio $\mathrm{Ca_{in}/Ca_{out}}$ is unity; the concentration ratio for Ca across the membrane is $\dfrac{[\mathrm{Ca}]_o}{[\mathrm{Ca}]_i} = \left(\dfrac{[\mathrm{Na}]_o}{[\mathrm{Na}]_i}\right)^r \exp - \left((r-2)\dfrac{FE_m}{\mathrm{RT}}\right)$.

Ideal Na/Ca Countertransport With ATP

In this system, all the thermodynamic rules listed above are obeyed. The effect of ATP is to facilitate or lubricate the carrier. Thus it can increase an influx and an efflux by a factor, *f*, but the flux ratio will remain unchanged since $\mathrm{Ca_{influx}}(f)/\mathrm{Ca_{efflux}}(f)$ is the same as above.

Diffusion of Ca

Measurements of $[\mathrm{Ca}]_i$ and a knowledge of $[\mathrm{Ca}]_o$ yield a value of +140 for E_{Ca}. With a membrane potential of −60, this yields a value of $zF(200)$ mV for the Ca electrochemical gradient. The flux ratio for diffusion is exp(400/25) or 8.9×10^6 (Ca_{influx}/Ca_{efflux}), and Ca influx in squid axons with $[Na]_i$ = 40 mM is 90% leak and perhaps 10% Na/Ca exchange even in the presence of ATP (12).

The Real Axon

The above considerations suggest that the carrier flux ratio for a squid axon is about 10 (efflux/influx), while the diffusional flux ratio (efflux/influx) is $1/8.9 \times 10^6$. The real axon, therefore, has the flux (fmole/cm^2) characteristics shown in Table 4.2 if ATP is considered to increase by 10-fold unidirectional fluxes. Note that the usual effect of increases in $[Na]_i$ is to promote Ca influx and decrease Ca efflux, so that going from a $[Na]_i$ of 40, as assumed above, to 80 would have the effects shown in Table 4.3. This table is constructed on the assumption that Ca influx by carrier is increased as the square of $[Na]_i$, while Ca efflux is decreased to half by a 40 mM increase in $[Na]_i$. This inhibitory action of Na_i is largely overcome by ATP. The point of Table 4.3 is that if we take total flux ratios, they have changed from 1.0 to about 1.3, but the carrier flux ratio (in the absence of ATP) has changed from 10 to 1.3 by the single change of $[Na]_i$. If one examines Ca influx, this has changed from 0.3 to 3 by the addition of ATP, or a 10-fold change (carrier flux). The experimentally measured flux changes

TABLE 4.2.

	Diffusion	Flux by carrier	Carrier + ATP	Total
Influx	27	0.3	3	30
Efflux	0	3.0	30	30

TABLE 4.3.

	Diffusion	Flux by carrier	Carrier + ATP	Total
Influx	27	1.2	12	40
Efflux	0	1.5	30	32

from 27 to 30 on the addition of ATP, since most is by diffusion. By contrast, Ca efflux is changed 10-fold by the addition of ATP.

Equilibrium Levels of $[Ca]_i$

With the apparent indication that the carrier flux ratio in squid axons is 10, one may ask what would be the equilibrium value of $[Ca]_i$ in a squid axon if there were no Ca leak. With $E_{Na} = 60$ and $E_m = -60$, we have Table 4.4.

We find that for a $[Ca]_o$ of 3 mM and a value of $r = 3$, the equilibrium value of $[Ca]_i$ is 200 nM; for $r = 4$, the equilibrium value of $[Ca]_i$ is 1.7 nM. This is not very different from the assumptions made earlier that $[Ca]_i$ is 20 nM in the working axon, and the carrier flux ratio is 10, a statement equivalent to saying that $[Ca]_i$ is 10 times the equilibrium value (i.e., 2 nM). This produces the appropriate flux ratio.

Na-free Ca Efflux

The definition of an uncoupled Ca efflux is that observed in the absence of Na_o. Experience with the Na pump has taught us that Li, Rb, Cs, and indeed all cations have an ability to promote Na pump flux; hence it is reasonable to assume that all cations have an ability to promote Ca extrusion. Experiments with aequorin show that Li reduces $[Ca]_i$ when this has been increased by choline sea water. Even choline has some ability to act as Na. If the $[Ca]_i$ is low, there is a slow rate of carrier cycling, and virtually any cation is capable of delivering enough carrier to the inside of the axon to promote Ca extrusion.

TABLE 4.4.

r	$r(E_{Na} - E_m)$	$2(E_{Ca} - E_m)$	E_{Ca}	$[Ca]_i$ (μM)
2	240	240	60	25
3	360	360	120	0.20
4	480	480	180	0.00167

KINETICS OF Na/Ca TRANSPORT

Unlike the thermodynamics of transport, where the energy requirements can be specified without regard to details, the kinetics of Na/Ca transport can only be related to a specific transport model. If the number of ions transported per cycle is large, then complexity of a high order is to be expected. The experimental observations in Table 4.5 in squid axons are related to inferences regarding possible models.

TABLE 4.5.

Observation	Model inference
1. Ca efflux versus $[Ca]_i$ hyperbolic	A single Ca is transported per cycle
2. Ca efflux versus $[Na]_o$ is S-shaped	Multiple Na binding outside per cycle of Na/Ca carrier
3. Ca influx versus $[Na]_i$ is S-shaped	Multiple Na binding inside per cycle of the Na/Ca carrier
4. Ca efflux versus $[Na]_i$, a decrease independent of absolute value of $[Ca]_i$	$[Ca]_i$ and $[Na]_i$ do not compete for a site; instead, a $[Na]_i$ increase forces more carrier to Ca influx mode (see 3)
5. Ca efflux increased by increases in E_m and decreased by depolarization	Loaded carrier has net charge such that more Na charges enter than Ca charges leave per cycle
6. $[Ca]_o/[Ca]_i = 10^5$	4 or more Na must be coupled to the movement of 1 Ca.
7. Ca efflux is maximal if $[Na]_i$ is 0	Carrier is capable of the reaction $(Na_o)_n XYCa_i \rightarrow (Na_i)_n XYCa_o \rightarrow {}_n Na_i + Ca_o + X_i$. $X_i \rightarrow X_o$; that is, the unloaded carrier can return to continue Ca efflux in the absence of Ca influx via carrier; alternately, that carrier will move either Na or Ca (competition) but conflicts with (4) above (X is Na-binding and Y is Ca-binding site)

The literature involving Na/Ca exchange deals with two different sorts of movement: in one case with a competitive interaction between a carrier R and Na and Ca such that the only carrier reactions are:

$$2\ \mathrm{Na} + \mathrm{R} = \mathrm{Na_2R}$$

$$\mathrm{Ca} + \mathrm{R} = \mathrm{CaR}$$

and there are two equilibrium constants for the interaction (K_{Na}, K_{Ca}). This is effectively Na/Ca countertransport, since the only net flux possible is when Na moves in one direction and Ca in the other. Since both Ca and Na are substantially more concentrated outside than inside a fiber, this is a system for moving both ions down their concentration gradients and has been frequently invoked to explain how more Ca enters a cell when $[\mathrm{Na}]_o$ is reduced.

The model developed by Benninger et al. (6) is as follows:

$$\begin{array}{ccccccc}
 & \text{Out} & & \text{Membrane} & & \text{In} & \\
\mathrm{Ca} \leftarrow & \mathrm{CaR} & \longleftarrow & \mathrm{CaR} & \longleftarrow & \mathrm{CaR} & \\
 & \downarrow & & & & \uparrow & \\
2\ \mathrm{Na} \rightarrow & \mathrm{R}^{-2} & & & & \mathrm{R}^{-2} & \leftarrow \quad \mathrm{Ca} \\
 & \downarrow & & & & \uparrow & \\
 & \mathrm{Na_2R} & \longrightarrow & \mathrm{Na_2R} & \longrightarrow & \mathrm{Na_2R} & \rightarrow \ 2\ \mathrm{Na}
\end{array}$$

The equilibrium reactions with the carrier are:

$$\frac{[\mathrm{Ca}]_o[\mathrm{R}]_o}{[\mathrm{CaR}]_o} = K_1 \qquad \frac{[\mathrm{Na}]_o^2[\mathrm{R}]_o}{[\mathrm{Na_2R}]} = K_2$$

$$\frac{[\mathrm{Ca}]_i[\mathrm{R}]_i}{[\mathrm{CaR}]_i} = K_3 \qquad \frac{[\mathrm{Na}]_i^2[\mathrm{R}]_i}{[\mathrm{Na_2R}]_i} = K_4$$

Elimination of $[\mathrm{R}]_o$ and $[\mathrm{R}]_i$ yields

$$\frac{[\mathrm{CaR}]_o}{[\mathrm{Na_2R}]_o} = \frac{[\mathrm{Ca}]_o}{[\mathrm{Na}]_o^2}\ \frac{K_2}{K_1}$$

$$\frac{[\mathrm{CaR}]_i}{[\mathrm{Na_2R}]_i} = \frac{[\mathrm{Ca}]_i}{[\mathrm{Na}]_i^2}\ \frac{K_4}{K_3}$$

so that for 0, net charge of the loaded carrier

$$\frac{[\mathrm{Ca}]_o[\mathrm{Na}]_i^2}{[\mathrm{Ca}]_i[\mathrm{Na}]_o^2} = \frac{K_1 K_4}{K_2 K_3}$$

If the carrier were such that Ca binding on both sides of the membrane is identical, then $K_1 = K_3$; if Na binding were also symmetical, then $K_2 = K_4$, so that the expression above reduces to

$$\frac{[\mathrm{Ca}]_o[\mathrm{Na}]_i^2}{[\mathrm{Ca}]_i[\mathrm{Na}]_o^2} = 1 \quad \text{or} \quad \frac{[\mathrm{Na}]_o^2}{[\mathrm{Na}]_i^2} = \frac{[\mathrm{Ca}]_o}{[\mathrm{Ca}]_i}$$

As the authors recognize, this would yield a $[\mathrm{Ca}]_i$ much too high. Asymmetry of the equilibrium constants is proposed, such that the affinity for Ca_i is high compared with that for Ca_o. This arrangement can only be made by an input of energy to the system. It is proposed that ATP produces this affinity change. A difficulty with such a proposal is, that when ATP is substantially reduced in a nerve fiber, the value for $[\mathrm{Ca}]_i$ is unaffected. Moreover, when Ca content is increased, the fiber recovers its original Ca content when ATP has been reduced to 1/100 of its physiological value. A final feature of this model is the proposal that the strong asymmetry of K_1K_4/K_2K_3 is reduced when the membrane is depolarized so that Ca entry is favored.

A modification of this scheme is to suppose that there is a carrier, X,[2] that can combine with either 2 Na or 1 Ca. If it combines with either, it can move to the inside of the fiber where it can discharge its ion and combine with either Na or Ca on the inside and then move outward and discharge the ion carried from inside to the outside. If we let A be either 2 Na or 1 Ca, then

$$\mathrm{A}_o + \mathrm{X}_o \leftrightarrows \mathrm{AX}_o \leftrightarrows \mathrm{AX}_i \leftrightarrows \mathrm{A}_i + \mathrm{X}_i$$

$$\mathrm{A}_i + \mathrm{X}_i \leftrightarrows \mathrm{AX}_i \leftrightarrows \mathrm{AX}_o \leftrightarrows \mathrm{A}_o + \mathrm{X}_o$$

[2]The notation X is for carriers where the unloaded form can also move; R is for nonmobile, unloaded carriers.

If the carrier can produce a net ion flux, as the Na/Ca system clearly can, then a free movement of unloaded carrier is necessary: $X_o \leftrightarrows X_i$.

Such a scheme has been proposed by Miller and Moicescu (14), who considered both a 2 Na/1 Ca and a 4 Na/2 Ca stoichiometry. A detailed analysis shows the following:

$$m_i^{Ca} = \frac{PK_o^{Ca}[Ca]_o^p\,[X]_T}{2(1 + K_o^{Ca}[Ca]_o^p + K_o^{Na}[Na]_o^{2p})}$$

where P is the rate of transfer of carrier, p is either 1 or 2, and $[X]_T$ is the total carrier concentration, while m_i^{Ca} is a Ca influx. The equation is remarkable in that it contains no term in $[Na]_i$, a parameter long recognized as critical in determining Ca influx.

Another feature of this model is the absence of enforced coupling between the fluxes of Na and Ca. It is possible, for example, for Na_o to combine with the carrier

$$2pNa_o + X_o \leftrightarrows Na_{2p}X_o \leftrightarrows Na_{2p}X_i \leftrightarrows 2pNa_i + X_i$$

$$X_i \leftrightarrows X_o$$

and the empty carrier return. This, then, is a mechanism for dissipating the Na gradient. Similarly, the Ca gradient would be rapidly dissipated by the arrangement proposed by a reaction sequence identical to that outlined for Na above.

A conclusion about carrier schemes is that they must be designed so that energy dissipation occurs only when an interaction with the ions to be transported is assured. Stated differently, if Ca is to be transported, its binding to the carrier should be the step that signals carrier translocation.

To return to the carrier reaction originally discussed:

$$2\ Na + R \leftrightarrows Na_2R$$

$$Ca + R \leftrightarrows CaR$$

The concentrations of Na and Ca on the two sides of the membrane are:

Ion	Outside	Inside
Ca^{2+}	2 mM	20 nM
Na^{+}	140 mM	10 mM
Ratio:		
Na/Ca	70	500,000

A change in affinity is needed for Ca versus Na on the inside that is high but that also alters when $[Na]_o$ is changed. Such a change would be necessary if the results of low-$[Na]_o$ solutions are to be explained. There is clearly a large Ca influx (and a contraction of cardiac cells). If Ca affinity inside were high, one would expect the carrier to retain Ca rather than to dissociate it and increase $[Ca]_i$.

A final observation (11) is that changes in $[Na]_i$ in squid axons alter the apparent activation constant of Ca efflux by $[Na]_o$. These findings can be understood by supposing that there are separate sites on the carrier for the binding of Na and Ca, and that translocation occurs when both Na and Ca binding sites are filled. Many observations require separate binding sites for Na and Ca on a single carrier, so that coupling of Na and Ca fluxes is enforced by moving an entity, Na_oXYCa_i or Na_iXYCa_o, where X is the Na binding site and Y that for Ca. Several questions must be considered in constructing such a model: (a) the stoichiometry Na/Ca, (b) the reaction sequence for Na and Ca binding, (c) the translocation step, and (d) the dissociation of the bound ions.

The energetics of Na/Ca transport strongly suggest a minimum of 4 Na/Ca. This value may be used provisionally in a transport model. Experimental information supports this value for stoichiometry. With respect to reaction sequence, some means must be found to ensure that the Na or Ca electrochemical gradients are not dissipated without countertransport. The carrier must load both Na and Ca before translocation can occur. From the point of view of loading, there is no reason to choose from first binding Ca or Na, followed by the other cation. From the effects of Na on carrier distribution between the two membrane surfaces, Na probably is bound first. This is inferred from the following considerations: (a) Ca + Y → CaY would mean that Ca is bound first, while (b) Na + X → NaX would mean that Na is bound first. The reaction sequence should then be

$$4\ Na_o + X_o \leftrightarrows (Na_4X)_o$$

$$(Na_4X)_o \rightarrow Na_4X_oY_i$$

$$Na_4XY_i + Ca_i \leftrightarrows Na_4X_oYCa_i$$

so that a binding of Na_o and Ca_i occurs before translocation. Because four charges (4 Na) move inward and two charges move outward (1 Ca) per cycle, there is a net transfer of 2+ inward when 1 Ca is translocated outward. The membrane potential thus assists in the movement of Na (in) and Ca (out). The next reaction after translocation is the dissociation reaction. Here Na presumably dissociates; the Ca binding site then is abolished so that Ca dissociates. The details of this reaction are of some importance since it is the binding of the fourth Na that induces a Ca binding site. It is the loss of the last Na from the carrier that allows the release of Ca. Because a number of other carrier unloading schemes are possible, the actual details of the unloading reaction remain unclear.

The thermodynamic relationship among [Na], [Ca], and E_m for $r = 4$ is:

$$\frac{([Na]_o)^4}{([Na]_i)^4} \exp\left(-\frac{2EF}{RT}\right) = \frac{[Ca_o]}{[Ca]_i}$$

A symmetrical rearrangement yields

$$([Na]_o)^4[Ca]_i \quad \exp(-E_mF/RT) = ([Na]_i)^4[Ca]_o \quad \exp(E_mF/RT)$$

The left-hand term is linear in $[Ca]_i$ and fourth power in $[Na]_o$, while it has an exponential dependence on E_m such that Ca efflux would increase with increases in E_m. It is possible to identify this expression with that for Ca efflux and right-hand expression with Ca influx.

Note that the expression contains a cross product, ($[Na]_o[Ca]_i$) or ($[Na]_i[Ca]_o$), which is of the right form for countertransport. The potential dependence is such that increases in E_m increase efflux and decrease influx. If we suppose that 4 Na^+ must combine with a carrier, X, in order that the carrier develop a Ca-binding site, Y, we have

$$4\ Na^+ + X \leftrightarrows Na_4X \leftrightarrows Na_4XY$$

$$Na_4XY + Ca \leftrightarrows Na_4XYCa$$

If Na_4XYCa has Na on the outside and Ca on the inside of the membrane, then E_m will assist the translocation of the carrier (given the usual inside negative values for E_m).

If, on the other hand, Ca is bound on the outside and Na on the inside, the membrane potential will retard translocation. Thus at equilibrium we can identify

$$Ca_{efflux} = K[Na_4X_oY_iCa]\ \exp\ (-\ EF/RT)$$

$$Ca_{influx} = K'[Na_4X_iY_oCa]\ \exp\ (EF/RT)$$

The values of K, K', will depend on the extent to which Ca efflux is activated by either $[Na]_o$, $[Na]_i$, and/or $[ATP]_i$. To understand Na activation, it is convenient to consider the carrier in the total absence of Ca and with Na confined to one side of the membrane. Since the carrier can move unloaded

$$[X]_o = [X]_i$$

As $[Na]_o$ is increased, NaX_o is formed, $[X]_o$ decreases, and $[X]_i$ decreases.

Clearly, we are forming NaX,. . ., Na_4X at the expense of $[X]_o$ and $[X]_i$. The addition of Na to one side of the membrane stimulates the removal of Ca from the other and inhibits the movement of Ca in the opposite direction since carrier otherwise available has been preempted by the large [Na] on one side.

These considerations show that apart from the effect of Na on the electrochemical gradient, there is a separate effect on the activation of influx and efflux of Ca. This movement has been addressed experimentally by Requena (11), who worked at a constant Na ratio of 5.5 in dialyzed squid axons. He was able to show that

27.5/5 Na_o/Na_i yielded a small Ca efflux, 110/20 mM yielded a maximal Ca efflux, while 440/80 mM yielded a smaller Ca efflux. These findings may be related to the activating effect of Na_o on Ca efflux and the inhibiting effect of Na_i on Ca efflux. Note that E_{Na} is constant but that what is being measured is a nonequilibrium flux.

Quantitative considerations show that inhibitory and activating effects can be explained by the relative amounts of carrier that are attracted to one membrane surface or the other by the binding of Na. A detailed study of this effect (17) shows that equations for Ca fluxes at equilibrium can be written as follows:

$$m_i^{Ca} = \frac{k_{-7}\,\beta[\mathrm{Na}]_i^4[\mathrm{Ca}]_o[\mathrm{X}]_T}{\alpha_o + \alpha_i + \beta([\mathrm{Na}]_o^4[\mathrm{Ca}]_i + [\mathrm{Na}]_i^4[\mathrm{Ca}]_o)} \quad [1]$$

$$m_o^{Ca} = \frac{k_7\beta[\mathrm{Na}]_o^4[\mathrm{Ca}]_i[\mathrm{X}]_T}{\alpha_o + \alpha_i + \beta([\mathrm{Na}]_o^4[\mathrm{Ca}]_i + [\mathrm{Na}]_i^4[\mathrm{Ca}]_o)} \quad [2]$$

where k_7 is exp $(-EF/RT)$, β is the product of all the rate constants for Na and Ca binding, $[X]_T$ is total carrier, $\alpha_o + \alpha_i$ is a summation of terms ($K_{Na}/[Na] = n$), such as $1 + n + n^2 + n^3 + n^4$ applied to $Na_o(\alpha_o)$ and Na_i (α_i), and k_{-7} is exp (EF/RT). (A detailed derivation is given in ref. 17.) These equations show that in addition to the term $[Na]^4[Ca]$ and one in membrane potential, there is a summation of terms in the denominator in [Na] on both sides of the membrane determines the partitioning of the carrier between the two membrane surfaces. A high value of [Na] on one side of the membrane not only increases the activation of the flux of Ca coming

→

FIG. 4.3. Ca efflux plotted in an upward direction and Ca influx downward on a semilog scale, each from a base value of 0.01 pmoles/cm^2 sec as related to the parameter $2E_{Na} - E_{Ca} - E_m$ (the net energy difference between the Na and Ca gradients). *Solid line,* slope of e-fold change in flux per 25 mV change in the parameter listed above. *Arrows,* at a value of zero difference in gradient, are at a value of influx and efflux of 30 fmoles/cm^2 sec. Points on the curve relate to changes in E_{Na}, E_{Ca}, and E_m. Flux ratios are 1, 2, or 3 for the three decades of fluxes plotted. Large values of Ca efflux depolarize, while large values of influx hyperpolarize, the membrane.

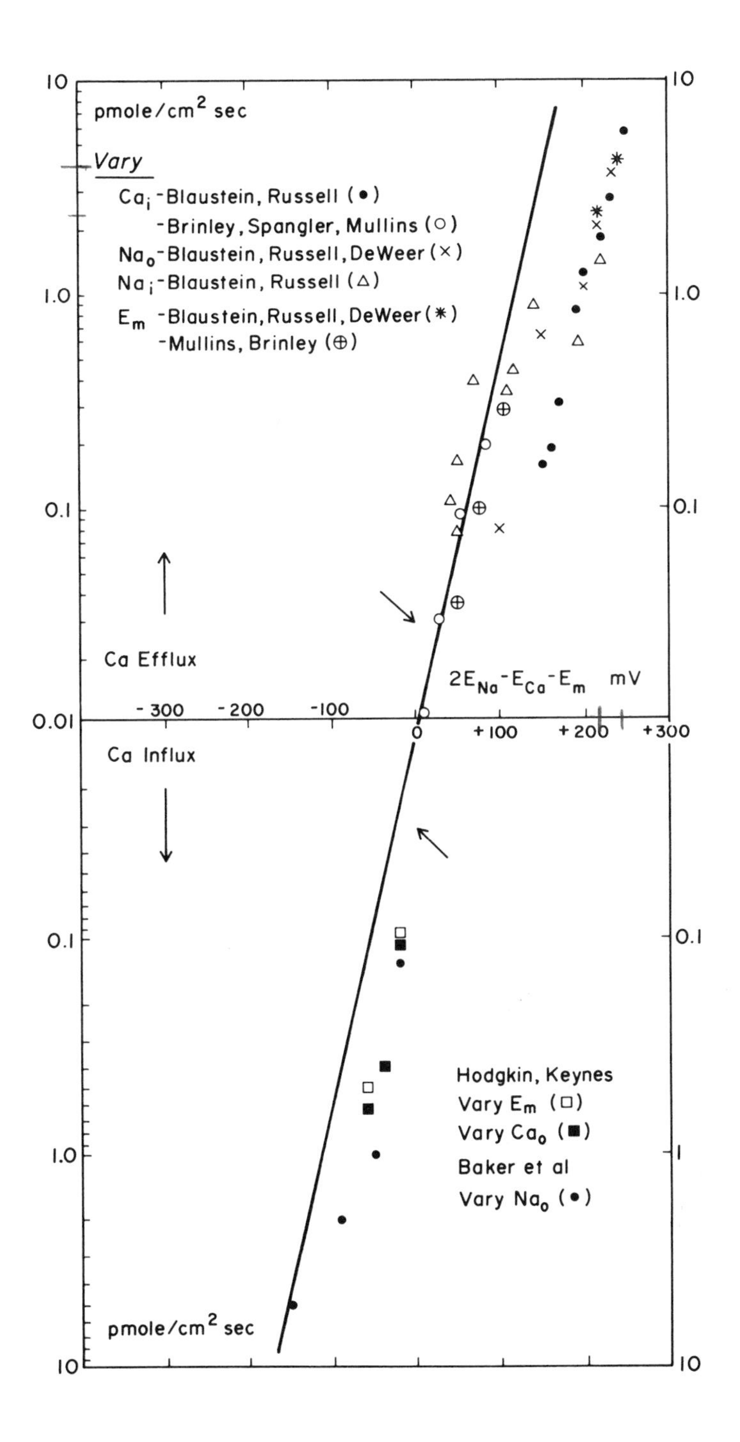
pmole/cm² sec
Vary
Caᵢ -Blaustein, Russell (●)
-Brinley, Spangler, Mullins (○)
Naₒ-Blaustein, Russell, DeWeer (×)
Naᵢ -Blaustein, Russell (△)
Eₘ -Blaustein, Russell, DeWeer (✱)
-Mullins, Brinley (⊕)
Ca Efflux
Ca Influx
2E_Na - E_Ca - E_m mV
-300
-200
-100
0
+100
+200
+300
10
1.0
0.1
0.01
Hodgkin, Keynes
Vary Eₘ (□)
Vary Caₒ (■)
Baker et al
Vary Naₒ (●)
pmole/cm² sec
1

from the other side of the membrane but also, by partitioning of the carrier, suppresses the flux of Ca moving from the opposite direction. Note that E_{Na} is the same for the same ratio of [Na] across the membrane (and $2E_{Na} - E_{Ca}$) regardless of the absolute concentrations of Na. Absolute Na concentration, however, is of critical importance in determining the magnitude of the Ca flux.

The equation above has been developed for the case of equilibrium. When the Na/Ca system is not at equilibrium, a variety of schemes for equations can be developed. Perhaps the most interesting case of net flux is that produced by changing E_m. In this case, using the admitted approximation of the equilibrium case, we have

$$I_C = (m_i - m_o)F$$

$$= \frac{(\exp(-EF/RT)\,[\mathrm{Na}]_i^4[\mathrm{Ca}]_o - \exp(EF/RT)\,[\mathrm{Na}]_o^4[\mathrm{Ca}]_i)\,X_T}{D}$$

where D is the denominator in equation [2]. This equation is of the form of the difference between two exponentials (or as sinh). This approximation is used throughout the following treatment. Clearly, when carrier loading rather than translocation becomes a rate-limiting step, the sinh relationship will fail and current will become potential independent. Another way of expressing the current-generating ability of Na/Ca exchange is to note that $I_C = 0$ when $E_m = E_R$ and that $E_R = 2E_{Na} - E_{Ca}$ and $I_C = k \sinh(E_R - E_m)$, at least for small displacements.

An empirical observation is that Ca fluxes are large when $E_{Na} - E_{Ca}$ is large compared with E_R. This point is illustrated in Fig. 4.3, where Ca efflux and Ca influx measurements from a variety of sources are plotted against $(2E_{Na} - E_{Ca} - E_m)$. The measurements summarize changes in $[\mathrm{Na}]_o$, $[\mathrm{Na}]_i$, $[\mathrm{Ca}]_o$, $[\mathrm{Ca}]_i$, and E_m on both Ca influx and efflux. They suggest that large Ca fluxes are a consequence of a large driving force on the Na/Ca exchange system. At equilibrium, the fluxes are many orders of magnitude smaller. The plot also shows that a change from a net inward to a net outward flux occurs at values predictable from thermodynamic considerations.

REFERENCES

1. Wilbrandt, W., and Koller, H. (1948): *Helv. Physiol. Pharmacol. Acta*, 6:208–211.
2. Lüttgau, H. C., and Niedergerke, R. (1958): *J. Physiol. (Lond.)*, 143:486–505.
3. Reuter, H., and Seitz, N. (1968): *J. Physiol. (Lond.)*, 195:451–470.
4. Blaustein, M. P., and Hodgkin, A. L. (1969): *J. Physiol. (Lond.)*, 200:497–527.
5. Baker, P. F., Blaustein, M. P., Hodgkin, A. L., and Steinhardt, R. A. (1969): *J. Physiol. (Lond.)*, 200:431–458.
6. Benninger, C., Einwachter, H. M., Haas, H. G., and Kern, R. (1976): *J. Physiol. (Lond.)*, 259:617–645.
7. DiPolo, R., and Beauge, L. (1979): *Nature*, 278:271–273.
8. Requena, J., DiPolo, R., Brinley, F. J., Jr., and Mullins, L. J. (1977): *J. Gen. Physiol.*, 70:329–353.
9. Requena, J., Mullins, L. J., and Brinley, F. J., Jr. (1979): *J. Gen. Physiol.*, 73:327–342.
10. DiPolo, R. (1976): *Fed.Proc.*, 35:2579–2582.
11. Requena, J. (1978): *J. Gen. Physiol.*, 72:443–470.
12. DiPolo, R. (1979): *J. Gen. Physiol.*, 73:91–115.
13. Requena, J., and Mullins, L. J. (1979): *Quart. Rev. Biophys.*, 12:371–460.
14. Miller, D. J., and Moicescu, D. G. (1976): *J. Physiol. (Lond.)*, 259:283–308.
15. Horakova, M., and Vassort, G. (1979): *J. Gen. Physiol.*, 73:403–424.
16. Jundt, J., Porzig, H., Reuter, H., and Stucki, J. W. (1975): *J. Physiol. (Lond.)*, 246:229–253.
17. Mullins, L. J. (1977): *J. Gen. Physiol.*, 70:681–695.

Chapter 5

Measurement of Transport in Cardiac Cells

INTRODUCTION

Many conclusions in the literature of cardiac electrophysiology were reached at times when adequate methods were not available for the measurements that were undertaken. The following sections have been included to emphasize that while progress continues to be made on a variety of fronts, the small size and anatomic complexity of cardiac fibers still pose formidable challenges to the experimenter. The treatment given is critical of many of the measurements that have been made. Methods that employ fast flow of experimental solutions and small fiber bundles are likely to yield worthwhile results; the perfusion of whole hearts or the use of large isolated pieces of tissue are guaranteed to provide only further confusion.

A detailed consideration of electrical measurements, and voltage clamp in particular, is avoided in favor of dealing with transport. It is useful to contrast the measurement of cardiac action potentials with a microelectrode with the measurement of an isotopic flux on a piece of cardiac tissue. In the first instance, one is dealing with a measurement on what amounts to a single cell on the surface of a tissue; if care has been taken to make flows in the experimental bath adequate, then solution changes can be effected in less than 1 sec. In the case of isotopes, a reasonably large piece of tissue is necessary if enough counts are to be collected for measurement. A population of cells then is involved, and an extracellular space and all the complexities that accompany compartmental analysis. Solutions cannot be changed quickly, regardless of how fast the bath flow, because it is the extracellular space that limits these changes. The development of methods for transport study should focus on those that can be applied to single cells or the nearest approximation thereto.

ISOTOPE FLUXES

A powerful tool in the analysis of transport of monovalent alkali cations has been the use of radioisotopes to measure unidirectional fluxes. In cells of reasonably large size, such measurements have allowed the experimenter to characterize, for example, Na influx and efflux and to relate these to either diffusional or carrier-mediated movements. Difficulties arise with flux measurement if (a) the cells are small, (b) the permeability of the ion to be measured is high, and (c) the cell population is a mixture of different types of cells or tissues.

Small cells have a large surface/volume ratio (S/V), which determines the rate at which isotopes equilibrate with cells. Measurements of the rate of change of $[Na]_i$ when K-free conditions are applied (1) show that the time constant for the change of Na^+ in Purkinje fibers is of the order of 1 min. For Ca^{2+}, in amphibian heart, it is Na/Ca exchange that removes the major share of Ca after a contraction; this loss of Ca has a time constant of the order of 50 to 100 msec. These time constants are equal to or substantially less than that by which substances can be removed from the extracellular space of cardiac muscle. If the equilibration of an extracellular space and that of a cell have similar time constants, it will not be possible to separate the fluxes from the two compartments.

The ion permeability and S/V determine the time constant of equilibration. In the case of cardiac Purkinje fibers, measurements with ion-sensitive microelectrodes and measurements of that fraction of the membrane potential generated by Na efflux both suggest that the time constant for Na equilibration across the membrane is, as noted above, of the order of 1 min, a time substantially less than that for the equilibration of a reasonably thick bundle of cardiac fibers. Clearly, measurements on a single cell impaled with an ion-specific microelectrode at the surface of a tissue bundle will give a more precise estimate of ion fluxes than will isotopes that must penetrate several layers of cells and must diffuse through tortuous extracellular space pathways.

One test of the adequacy of isotopic flux measurement is to load a tissue with isotope and then measure the washout of radioactivity as a function of time. If a constant flux is to be measured, the ratio (cpm/m ÷ total cpm) will be constant; this is the rate constant for the loss of radioactivity and can be transformed to units of flux by using the internal ion concentration and the S/V ratio. Extensive measurements have been made by Langer (2), who showed that in cardiac tissue, the rate constant declines continuously with time. This implies that concentrations in the fiber are changing or there are a number of compartments with different rate constants unloading, or both. A difficulty with this measurement is that it takes at least 1 hr for the rate constant to decline to even a moderate value; meanwhile, approximately 99% of the counts have been lost. It is difficult to be sure that the remaining efflux is from cardiac cells rather than other cells in the tissue.

With respect to Ca and other divalent cations, rather than Na and monovalent cations, the difficulties are multiplied. The immersion of tissue in ^{45}Ca solutions loads extracellular structures preferentially, and the time constant for Ca uptake is short. When efflux measurements are begun, it requires an even longer time to clear the extracellular space and structures of ^{45}Ca counts. Various strategies are used to overcome these extracellular space difficulties. One is to use La^{3+} to exchange for Ca and thus release extracellular ^{45}Ca more quickly. This leaves a residue of La, an inhibitor of Ca fluxes, and still does not clearly define a Ca efflux. Another technique is to use a Ca chelator, such as EGTA, to trap ^{45}Ca counts as they emerge. This makes $[Ca]_o$ effectively zero and hence produces a highly nonsteady state [Ca] intracellularly. It may also actually facilitate the efflux of Ca. Other procedures, such as using EGTA/CaEGTA buffer mixtures to define a [Ca] in tissue spaces, suffer from the difficulty that the [Ca] they can produce is low and hence must be expected to reduce cell Ca concentrations.

In summary, flux measurements, in the case of cardiac cells, must be approached with caution. Order of magnitude measurements are probably reliable, and values for Na and K fluxes are more likely to be accurate than are Ca fluxes.

Measured ^{45}Ca influxes into a variety of cardiac cells have been tabulated by Chapman (3) and compared with either voltage clamp I_{si} measurements or with the time derivative of the rise of the action potential in TTX-poisoned fibers. The Ca influx deduced from such measurements is between 0.2 and 2.0 pmoles/cm^2 beat and the calculated increase in $[Ca]_i$ between 1 and 3 μM. A difficulty with the isotope measurements is that the time constant for Ca movement across the cell membranes is less than 100 msec; thus a substantial correction for back flux must be made. For electrical measurements, Ca fluxes can be underestimated by failing to allow for carrier-mediated Ca currents.

ION-SENSITIVE ELECTRODES

The construction of a recessed tip, ion-sensitive microelectrode by Thomas (4) was a substantial technical advance in electrophysiology in the past decade. These electrodes for measuring intracellular pH or [Na] are reliable when introduced into large nerve cell bodies or into giant axons. As cell size declines, however, a number of difficulties arise that may limit the usefulness of such devices in cardiac cells. First, in small cells, the electrodes cannot be reduced in size proportionally; thus they make a larger hole in a smaller cell with consequences still to be analyzed. When a large electrode and a small electrode are introduced into a cell to measure, for example, membrane potential, the larger electrode will be expected to measure a lower value because the shunt or leak around its point of insertion is larger. An ion-sensitive electrode is of no use unless an independent measure of membrane potential is available; the electrode measures both membrane potential and a contribution from its ion sensitivity. In cardiac cells, it is impossible to introduce ion-sensing and potential measuring electrodes into the same fiber; one must rely on the syncytial properties of the tissue to correct for membrane potential. Finally, because it only recently proved possible to introduce ion-sensing electrodes into beating cells, most measurements of $[Na]_i$ are for quiescent cells.

A recent study (5) measured $[Na]_i$, tension, and Na pump current in Purkinje fibers under voltage clamp while they were driven at 0.1 Hz. During a 6-min Rb_o-free test, $[Na]_i$ doubled and tension

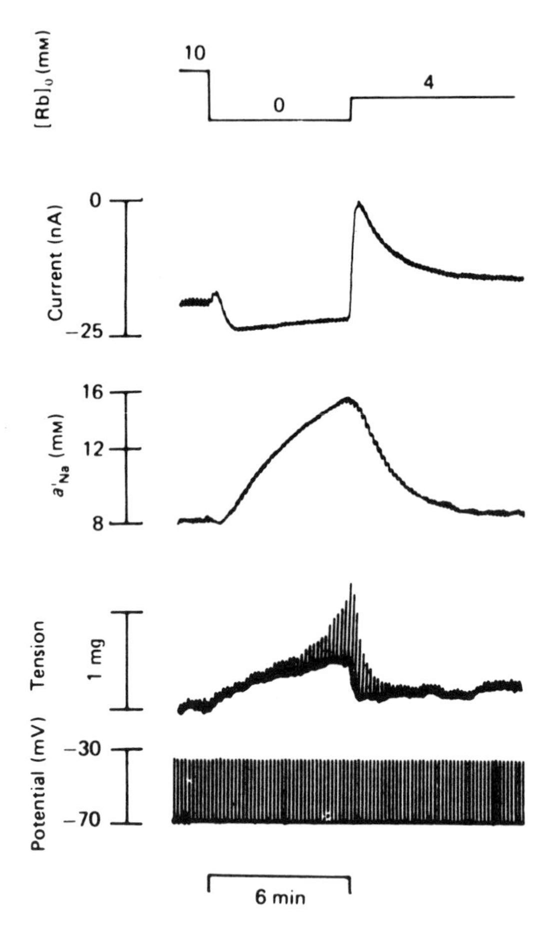

FIG. 5.1. Simultaneous measurement of membrane current, a^i_{Na}, tension, and voltage-clamped membrane potential. The holding potential was −68 mV. A 500-msec depolarizing pulse to −34 mV was applied at 0.1 Hz. The current and a^i_{Na} records have been filtered (time constant 3 sec). (From ref. 5.)

(both with depolarization and in its absence) rose many fold, indicating a clear connection between $[Na]_i$ and tension. The results are shown in Fig. 5.1.

Despite the foregoing reservations, measurements that have been made probably reflect accurately changes in $[Na]_i$, although the absolute values may be in some error. Nevertheless, ion-specific electrodes likely represent the best chance for measuring $[Na]_i$ continuously. The high impedance of present electrodes limits the response time severely; still one does not expect $[Na]_i$ to alter during the course of a single action potential, since the maximal entry (50 pmoles/cm^2 imp) would increase $[Na]_i$ by 0.02 mM.

ANALYTIC MEASUREMENTS

Numerous measurements have been made of the ion content of heart; like all such measurements, these must be related to cellular ion contents on the basis of a correction for extracellular space. Substantial difficulties exist in reaching a conclusion about the amounts of ions, such as Na^+ and Ca^+, which are primarily extracellular, that are inside cells. Small errors in the estimate of the extracellular fraction can make large errors in cellular estimates. For K, considerations of this sort are less of a problem, since the ion is primarily an intracellular constituent.

We can begin the construction of a table of analytic values for cardiac cells. Since there is little K in the extracellular space, an analytic value of 150 mM for internal cell K is reasonable. For Na, it is necessary to recognize that if cardiac cells are not beating, $[Na]_i$ may have a value different from that when bioelectric activity is being carried out. It is difficult to make the extracellular space corrections with any accuracy; bioelectric measurements may be more useful. In a variety of cardiac cell types, the peak of the action potential reaches +30 mV as a maximum, a value less than that reached by a normal squid axon. At the same time, the cardiac cell is protected from delayed rectification by the fact that a majority of its K channels are of a different type; E_{Na} is likely not greater in cardiac cells than in skeletal muscle or nerve fibers. In both these cell types, the Na ratio across the cell membrane is 11, or E_{Na} = 60 mV. In the case of cardiac cells with a constant Na load as a result of Na entry to effect both depolarization and Ca efflux, steady state $[Na]_i$ is even higher than in those tissues in which bioelectric activity is less continuous. As a matter of standardizing calculations, $[Na]_i$ is assumed to be 20 mM in working cardiac fibers; the Na ratio is 7, or E_{Na} = 50 mV. Measurements have been made of a^i_{Na} (activity inside) as low as 7 mM. With the usual activity coefficient of 0.75, this is 9.3 mM on the concentration scale, or a ratio of 15 across the cell membrane (1,6). This yields a value for E_{Na} of +70 mV. These measurements were made on resting fibers, however, and if E_{Na} remotely approached such large values, it is surprising that the action potential does not reflect it.

Time constants (that is, the time required for an ion concentration to rise or fall *e*-fold, where *e* is the base of natural logs) are useful in understanding flux measurements. For example, if $[Na]_i$ is assumed to be 10 mM (10 μmoles/cm^3), the time constant for the change in $[Na]_i$ to be 1 min, and the V/S ratio for cardiac fibers taken as 2.5×10^{-4} cm, then the product—concentration (1/time constant, V/S)—is the flux necessary to produce a *e*-fold change in concentration in the time specified; in this example, it is 42 pmoles/cm^2, a value close to that measured in cells undergoing bioelectric activity. By contrast, if we have a Ca flux of 10 pmoles/cm^2 (equivalent to 1 μA/cm^2 in measured Ca current) in the same preparation and $[Ca]_i$ is at a resting level of 50 nM (50×10^{-12}/moles/cm^3), the time constant for a change of $[Ca]_i$ is 1.3 msec. In a fiber undergoing contraction where $[Ca]_i$ is 1 μM, the time constant is 25 msec. These short time constants for Ca are necessary if contraction is to be brought about by transmembrane Ca flux. They preclude the measurement of Ca movement by isotope techniques, however, as they are many times faster than diffusion in the extracellular space.

What can be measured by isotope techniques in some instances is the time constant of Ca interchange between internal stores and the extracellular fluid. As an example of this, Wendt and Langer

(7) have shown that the time constant for the loss of Ca by cardiac cells in tissue culture is 24 min. This is increased to 33 min when the cells are bathed in Na-free solutions.

Analytic values for Ca are not useful in deciding on a value for $[Ca]_i$, since virtually all the cellular Ca is bound, complexed, or stored in intracellular compartments. Direct measurements of ionized Ca in nerve fibers using aequorin or arsenazo show values of 20 to 50 nM. In muscle fibers, aequorin signals are too small for resting values of Ca to be approximated. Direct analysis of cardiac tissue does show a value of Ca of about 1 mmole/kg. Most of this Ca is contained in extracellular compartments, such as connective tissue (2).

The fact that a cardiac fiber is relaxed means that $[Ca]_i$ must be 100 nM or less. The major recognized intracellular storage site for Ca is the SR. This organelle is capable of concentrating Ca by a factor of 10^4. Since its presence in cardiac cells is of the order of 1% of the volume, the capacity for Ca storage is 100 nM $\times$ 10^4 $\times$ 0.01 or 0.01 mM, a value that is 1% of total Ca. If we take this value (10^{-8} moles/cm^3) and multiply by 1/time constant, or 0.03 min^{-1} $\times$ 1/60 min/sec $\times$ 2.5 $\times$ 10^{-4}/cm, we get a Ca flux of 1.3 fmoles/cm^2. If a larger fraction of the cell Ca were participating, then the flux would be correspondingly larger. This is a negligible fraction of resting Ca flux. Mitochondria can store large quantities of Ca but appear to have a threshold for Ca uptake around 1 μM. Thus at rest they are unlikely to be an important storage site.

Under Na-free conditions, there is a large but transient Ca influx that ceases only when $[Na]_i$ reaches low levels. Busselen and Van Kerkhove (8) have shown that Na-free salines increase the Ca content of goldfish ventricles by 1 mmole/kg. Since Na-free conditions are unlikely to affect extracellular Ca binding, virtually all this extra Ca must be stored intracellularly, or such stores must have increased from 0.01 to 1 mM. This is a gain of 1 mmole in about 30 min or a flux of 10^{-6}/moles/cm^3 divided by 1,800 sec and multipled by 2.5 $\times$ 10^{-4}/cm or 138 fmoles/cm^2.

Wendt and Langer (7) also studied the arterially perfused interventricular septum and found gains of Ca of 0.2 mmoles/kg wet weight in 20 min when the preparation was treated with Na-free solutions. This lower Ca flux presumably represents the differing ability of the two tissue types to retain $[Na]_i$. Note that in both cases, the Ca gain is not continuous but occurs most rapidly in the earliest part of the exposure to Na-free conditions. A similar finding with respect to the development of tension in Na-free solutions is described below.

The anatomic complexity of cardiac tissue is substantial. In addition to internal compartments that can take up Ca, substances in sarcoplasm can complex Ca, and external compartments can also take up Ca. Not surprisingly, this myriad of compartments, as well as a variety of mechanisms for Ca transfer across the cell membrane, leads to a complex kinetics for the movement of ^{45}Ca from cardiac tissue. Indeed, Langer and Brady (9) have identified five compartments in the washout of ^{45}Ca from heart. Most workers with isotope efflux measurements find that a two-compartment-in-series model is the maximum degree of complexity that can be tolerated in the analysis of tracer kinetics; thus the continuously varying rate constant characteristic of ^{45}Ca washout in heart precludes detailed analysis. In addition, since the time constant for Ca movement across the cardiac cell membrane is substantially less than extracellular space washout, ^{45}Ca efflux that is measured almost certainly reflects primarily extracellular rather than intracellular Ca loss.

A scheme for compartments for Ca is shown in Fig. 5.2, from which it can be appreciated that other cells or connective tissue, or extracellular membranous structures, are likely contributors to ^{45}Ca efflux. Internal compartments 1, 2, and 3 might be identified with the mitochondria, the SR, and with other Ca-complexing entities. Each of these is in steady state with the $[Ca]_i$. Experimental methods selectively release Ca from at least compartments 1 and 2. The $[Ca]_i$ is maintained at a particular value by the balance of fluxes across the sarcoplasmic membrane generated by dissipative movements of Ca via channels and leaks and by Na/Ca exchange and

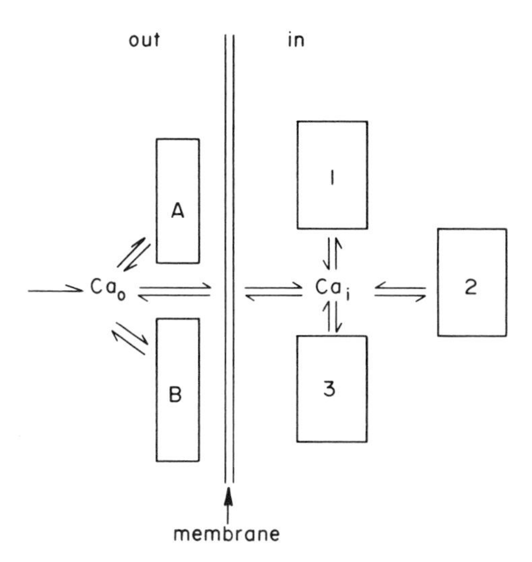

FIG. 5.2. Kinetic situation for Ca movement from outside to inside the cell. External compartments A and B operate to stabilize $[Ca]_o$. Internal compartments 1, 2, and 3 are, respectively, mitochondria, the SR, and other Ca-complexing entities.

possibly other Ca pumps. Externally, there is a similar situation, in which $[Ca]_o$ is in some steady state with external Ca reservoirs or buffers that tend to stabilize $[Ca]_o$. Washout kinetics for Ca involve a high order of complexity.

The foregoing considerations have dictated the construction of Table 5.1 for ion concentrations in cardiac cells. The table reflects the following considerations: (a) Tissue analytic K can be equated virtually directly with $[K]_i$; (b) E_{Na} in a working fiber is not more than 25 mV positive to the peak of the action potential; in a resting fiber, the Na pump may reduce Na_i to half (a value close to that found with ion-specific electrodes); and (c) the maximal $[Ca]_i$ compatible with a relaxed fiber is 100 nM; a reduction of $[Na]_i$ by twofold will produce a large $[(2)^4$-fold$]$ reduction in $[Ca]_i$. Since part of the $[Ca]_i$ is contributed by leaks, only a fivefold reduction has been made.

The values of E_{Na} and E_{Ca} in Table 5.1 mean that in a working

TABLE 5.1. *Ion concentrations in cardiac fibers during diastole*

Ion	Extracellular	Intracellular	
		Working fibers	Resting fibers
Na mM	140	20 (50)[a]	10 (66)
K mM	4	140 (−90)	150 (−91)
Ca nM	2×10^6	100 (125)	20 (145)

[a]Values in parentheses are ion equilibrium potentials in mV.

fiber, $E_R = -25$ mV; in a resting fiber, $E_R = -13$ mV. During maximal contractions, E_{Ca} may approach +60; thus resting and working values for E_R would be +72 and +40, i.e., values never approached by the action potential. On the other hand, disabling the Na pump so that $E_{Na} = +30$ would lead, at least transiently, to $E_R = -80$ mV. Thus, E_R can vary between −80 and +70 mV, so that the equilibrium potential for the carrier current can occur over the entire range of E_m reached by the action potential.

AEQUORIN AND ARSENAZO

Ion flux measurements of Ca, even in squid giant axons, have unsatisfactory aspects. In ^{45}Ca-injected squid axons, there is clear evidence that Na-free solutions so change the specific activity that the original baseline is lost (10). In dialyzed squid axons, the contamination of ordinary reagents with Ca makes necessary the use of CaEGTA buffers to control ionized Ca. This in turn leads to a loss of physiological control of $[Ca]_i$ and to some poorly understood effects that EGTA itself may have on the operation of Na/Ca exchange. For these reasons, the measurement of net Ca fluxes as changes in either the analytic or ionized [Ca] have proved indispensable in understanding certain aspects of Ca regulation.

With aequorin, a protein isolated from jellyfish, there is a sensitivity to Ca that begins at about 10 nM, is about linear with changes in Ca at 50 nM (11), becomes a 2.5 power function of [Ca] at about 1 μM, and subsides to zero at about 100 μM. Measurements with this material in freshly isolated squid axons show that $[Ca]_i$

is 20 to 50 nM (12), and that the material is a good detector of whether or not a particular experimental condition produces a net Ca flux (13). In particular, it has been possible to show that the destruction of ATP by apyrase injection does not lead to a net flux of Ca in the fiber. Aequorin has also been used in cardiac (14) and skeletal muscle (15) fibers to show that a light transient from aequorin-Ca interaction accompanies the measured mechanical response. Unfortunately, in neither tissue is there a "resting glow" that would allow the aequorin reaction to indicate $[Ca]_i$ when this is at its lowest value. This glow is a useful indicator as to whether the fiber is losing or gaining Ca in squid axons. Hence it can be inferred that in frog fibers, the resting $[Ca]_i$ is at least as low or lower than squid axons; otherwise, a light output would be measurable. A frog skeletal muscle fiber is about one-fifth the diameter of a squid axon and so would have 1/25 of the light emission if $[Ca]_i$ were the same in both fibers. A reduction of light emission in squid axons by this amount still yields measurable light (16). Because of the highly nonlinear response of aequorin to [Ca], it is useful principally at extremely low ionized Ca concentrations, where its sensitivity is indispensable to the measurements to be undertaken.

The dye arsenazo III has been used as a method of measurement of $[Ca]_i$ inside squid giant axons as well as in other nerve cells. The principle of its measurement is that Ca associates with the dye, and the Ca-arsenazo complex has a different absorption spectrum from that of the undissociated dye. This reaction can measure with great linearity Ca concentration changes well into the micromolar range, although the sensitivity of the arsenazo reaction may be inadequate at [Ca] below 50 nM. Refinements in spectrophotometry and in the introduction of dyes into cells may make this procedure practical for estimating the resting $[Ca]_i$ of cardiac fibers.

ELECTRICAL MEASUREMENT OF ION FLUX

As indicated above, Na/Ca transport is a highly reversible reaction that can move Ca in either direction, depending on whether the membrane potential is positive or negative to E_R. Another exciting development in cardiac electrophysiology is the discovery that there is a gated channel that carries slow inward current that in large measure represents inward Ca movement. There has been no consideration of the proposition that measurements of inward Ca current may be in serious error, because the behavior of carrier current has not been considered. This point, therefore, is discussed here.

If the carrier current of Na/Ca exchange, I_C, is considered not to be gated but to rise instantaneously with depolarization, it may be seen merely as one of the many background currents in the cardiac fiber and thus can be ignored in interpreting the time-dependent current records produced by the *si* current. Unfortunately, this interpretation is not valid; depolarization for even rather short periods of time must be expected to change E_{Ca}. Such a change affects I_C, which, unlike I_{Ca} (defined as the component of I_{si} that carries Ca), is highly sensitive to a change in E_{Ca}. The details of this change are considered in Chapter 7; it is sufficient here to note that the necessary experimental work has not been done to separate I_{Ca} from I_C. Measurements have been made of the reversal potential of I_{si}, and there is uniform agreement that this is far from E_{Ca} (often it is at E_{Na}). It has been suggested that this behavior is attributable to the flow of Na through the slow inward channels. An alternate explanation is that Na/Ca exchange generates a current that subtracts from that in the Ca channels; hence the reversal potential must be expected to lie between that for the Na/Ca E_R and E_{Ca}.

Much of the difficulty in reviewing the voltage clamp literature arises from the common assumption that there are only three relevant reversal potentials: (a) a highly negative E_K, (b) a moderately positive E_{Na}, and (c) a highly postive E_{Ca}. A correlate is that any current reversing at potentials less than zero would be K movement, while reversal potentials that are modestly positive would be either I_{Na} or I_{Ca} with a substantial contribution from Na. It is this flexibility in current-voltage diagram interpretation that increases doubts about the validity of much analysis. An impartial reviewer may claim that the burden of proof that voltage clamp data are generally valid lies with the investigators who propose such data. Early questions about voltage clamp data were raised by Johnson and Lieberman (17). A recent review by Beeler and McGuigan (18), while deploring the unfortunate language of the earlier review, suggests that great care should be used in accepting voltage clamp data as valid. Difficult questions, such as how one deals with a reversal potential that involves the flow of three separate ionic currents, are not mentioned. On the other hand, the authors recognize the difficulty of a reversal potential for I_{Ca} where, by hypothesis, Ca is still flowing inward, Na is flowing outward, and the net current is zero. If one adds a further current, namely that of I_C where Na is flowing outward and Ca inward (and current is outward), the further complexity of this situation can be appreciated. There is little inclination to deal with the difficulties of a rapidly changing E_{Ca} as a result of current flow or of an even more rapidly changing value for E_R.

ION SUBSTITUTION

The separation of Na and K currents in squid axon was possible because Na_o was replaced by choline so that the kinetics of the K current could be studied. Such ion substitution experiments in squid axons were successful because of the large size (and correspondingly small S/V ratio) of such axons. By comparison, the development of a strong Na/Ca exchange in cardiac cells means that Na-free conditions bring about a large Ca loading of the fiber and contraction is many fiber types. Ca movement across the cardiac cell membrane is so rapid that isotope experiments can only roughly approximate some of the slow movements of Ca into stores.

The replacement of Cl^- by other anions has been shown to affect $[Ca]_o$. Since Cl has been implicated in H^+ pumping, and such pumping is necessary if $pH_o = pH_i$, Cl removal may lead to significant changes in pH_i. Finally, a relationship exists between pH_i and $[Ca]_i$, such that decreasing pH_i increases $[Ca]_i$, and increasing pH_i has the opposite effect. This has been shown in barnacle muscle (19) and in squid giant axon (16).

One can conclude, therefore, that ion substitution experiments are of limited applicability to cardiac cells. Ca_o-free conditions lead to fibers that do not contract and to a depletion of internal Ca stores, whereas Na_o-free conditions lead to fibers that operate in ways that differ significantly from normal, because neither the introduction of Ca during depolarization nor the pumping out of such Ca is possible via Na/Ca exchange.

TRANSIENT NATURE OF Ca ENTRY

In squid giant axons, the removal of Na_o and its replacement by choline or Li leads to a large and sustained increase in Ca influx. An equivalent replacement in cardiac fibers of Na_o by other ions must be expected to lead only to a transient increase in Ca content since, as noted earlier, this reversal of the Na gradient can promote an increased Ca influx only as long as $[Na]_i$ remains finite.

The evidence from many measurements in cardiac fibers is that $[Na]_i$ declines with a time constant of the order of 1 min under Na_o-free conditions. Thus changes in $[Na]_o$ must be expected to lead to changes in $[Na]_i$; in the simplest case, a given decrease in $[Na]_o$ will lead to an equivalent change in $[Na]_i$, so that E_{Na} in the steady state will be unchanged. The practical consequences of this change are that Ca entry in response to a change in $[Na]_o$ will be a transient or step change in Ca entry, leading to a sustained increase in Ca content of the fiber.

If instead of a change in $[Na]_o$, E_m is changed to a new value, this change will lead in the steady state to a change in E_{Ca} of equal magnitude such that $2E_{Na} - E_{Ca} - E_m$ has the same value before and after the change in E_m. These changes in both $[Na]_o$ (as a change of E_{Na} of 25 mV) and E_m (a 25-mV hyperpolarization) are shown in Fig. 5.3, in which it can be seen that both changes, while leading to a Ca gain by the fiber, lead to a gain that is not continuous but transient. Both changes in Ca content are fully reversible once $[Na]_o$ or E_m is returned to its original value. The point is important, as it emphasizes that cardiac fibers cannot be expected to gain Ca continuously under these conditions.

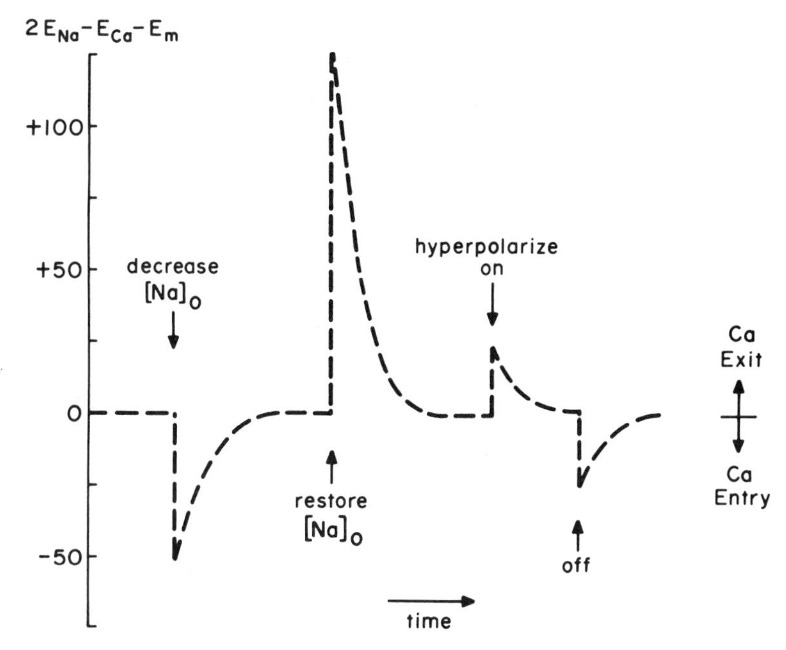

FIG. 5.3. Thermodynamic parameter ($2E_{Na} - E_{Ca} - E_m$) plotted as a function of time for two conditions: (a) $[Na]_o$ is decreased such that the initial change in E_{Na} is 25 mV. A subsequent decline in $[Na]_i$ is such that E_{Na} returns to its initial value. Restoration of $[Na]_o$ leads to a Ca exit until $[Na]_i$ returns to its original value. (b) Hyperpolarizing the membrane to 25 mV leads to a change in E_{Ca} such that $[Ca]_i$ declines. Depolarization has an opposite effect.

INHIBITORS

A method often advanced as an aid in identifying the ionic currents in cardiac fibers is the use of inhibitors. It is suggested that substances, such as Mn^{2+} and La^{3+}, as well as organic compounds, such as isoproveratril, are specific inhibitors of the Ca channel. This is not so; all these substances have effects on Na or K channel currents, as well as (at least in the case of Na/Ca exchange) on Ca carrier movements. It is essential to recognize that there are no known inhibitors that will differentiate between a Ca current carried by a Ca channel and a Ca movement via Na/Ca exchange. The highly specific inhibitor of Na/Ca exchange is a reduction of $[Na]_i$; this change can have profound effects on cardiac contractility. A corollary is that an increase in $[Na]_i$ is capable of greatly enhancing Ca entry and contractility during the cardiac cycle.

REFERENCES

1. Deitmer, J. W., and Ellis, D. (1978): *J. Physiol. (Lond.)*, 277:437–453.
2. Langer, G. A. (1964): *Circ. Res.*, 15:393–405.
3. Chapman, R. A. (1979): *Prog. Biophys. Mol. Biol.*, 35:1–52.
4. Thomas, R. C. (1970): *J. Physiol. (Lond.)*, 210:82–83.
5. Eisner, D., Lederer, W., and Vaughan-Jones, R. D. (1980): *J. Physiol. (Lond.)*, 300:42P.
6. Lee, C. O., and Fozzard, H. A. (1975): *J. Gen. Physiol.*, 65:695–708.
7. Wendt, I. R., and Langer, G. A. (1977): *J. Mol. Cell. Cardiol.*, 9:551–564.
8. Busselen, P., and Van Kerkhove, E. (1978): *J. Physiol. (Lond.)*, 282:263–283.
9. Langer, G. A., and Brady, A. J. (1963): *J. Gen. Physiol.*, 46:703–719.
10. Blaustein, M.P., and Hodgkin, A. L. (1969): *J. Physiol.*, 200:497–527.
11. Allen, D. G., Blinks, J. R., and Prendergast, F. G. (1977): *Science*, 195:996–998.
12. DiPolo, R., Requena, J., Brinley, F. J., Jr., Mullins, L. J., Scarpa, A., and Tiffert, T. (1976); *J. Gen. Physiol.*, 67:433–467.
13. Requena, J., DiPolo, R., Brinley, F. J., Jr., and Mullins, L. J. (1977): *J. Gen. Physiol.*, 70:329–353.
14. Allen, D. G., and Blinks, J. R. (1978): *Nature*, 273:509–513.
15. Blinks, J. R., Rudel, R., and Taylor, S. R. (1978): *J. Physiol. (Lond.)*, 277:291–323.
16. Mullins, L. J., and Requena, J. (1979): *J. Gen. Physiol.*, 74:393–413.
17. Johnson, E., and Lieberman, M. (1971): *Annu. Rev. Physiol.*, 37:479–532.
18. Beeler, G., and McGuigan, J. (1978): *Prog. Biophys. Mol. Biol.*, 34:219.
19. Lea, T. J., and Ashley, C. C. (1978): *Nature*, 275:236–238.

Chapter 6

Interactions Between Na/K and Na/Ca Pumps

INTRODUCTION

Interaction between the Ca and Na pumps is inevitable, given the existence of a common ion, Na. Complexity is further heightened by the fact that both pumps can extrude Na, and one pump is influenced by membrane potential. Progress in understanding these processes can be made by considering simple cases; complexity can then be added to whatever degree desired.

It is assumed first that a cell has the usual concentrations of ions on both sides of the membrane, but that Na can enter only via Na/Ca exchange. Steady-state considerations require that Ca inward leak equal pumped Ca efflux, and that the associated Na entry equal the exit of Na via the Na/K pump. These assumptions yield Table 6.1. In one cycle of the Na/Ca and Na/K pumps, the depolarizing effect of Na entry in exchange for Ca is not compensated by the hyperpolarizing effect of Na/K pumping; 2+ charges enter and only 1.33+ charges emerge, leaving a balance to be met by a leak of K^+. This leak can only occur if E_m is positive to E_K.

TABLE 6.1.

Entity	Coupling	In	Out	Net charge
Na/Ca pump	4:1	4 Na^+	1 Ca^{2+}	2+ (in)
Na/K pump	3:2	2.66 K^+	4 Na^+	1.33+ (out)
Leak	—	1 Ca^{2+}	2.66 K^+	0.66+ (out)

If one now allows a leak of Na inward, such that 2 Na^+ enter by diffusion per 4 Na entering via Na/Ca exchange (a total of 6 Na^+), we have Table 6.2. In this situation, the depolarizing effect of Na entering via the Na/Ca mechanism and by diffusion is such that it is compensated by a hyperpolarization generated by the Na/K mechanism. The membrane potential cannot be at E_K; although the Na/Ca and Na/K pumps are in charge balance, no consideration has been given to passive K influx. The usual value of $[K]_o$ is approximately 30 times less than $[Na]_o$ and P_K about 30 times greater than P_{Na}. Hence passive K influx will be the same as passive Na influx, so that in Table 6.3, the leak will include 2 K^+ (in) and 6 K (out). This is a passive flux ratio of three (efflux/influx), indicating a displacement of the membrane potential 27 mV below E_K. A final tabulation of ion movements by pump and leak is shown in Table 6.3. Increasing P_K will raise E_m, because it will make the flux ratio closer to 1. Increasing P_{Na} (at least initially) will also increase the membrane potential, since it will increase the electrogenic contribution to the membrane potential as compared with the depolarizing action of the Na/Ca system.

The foregoing considerations are satisfactory for a resting cardiac

TABLE 6.2.

Entity	Coupling	In	Out	Net charge
Na/Ca	4:1	4 Na^+	1 Ca^{2+}	2+ (in)
Na/K	3:2	4 K^+	6 Na^+	2+ (out)
Leak	—	1 Ca^{2+}		
		2 Na^+	4 K^+	0

TABLE 6.3.

Entity	Coupling	In	Out	Net charge
Na/Ca	4:1	4 Na^+	1 Ca^{2+}	2+ (in)
Na/K	3:2	4 K^+	6 Na^+	2+ (out)
Leak	—	1 Ca^{2+}		
		2 Na^+		
		2 K^+	6 K^+	0

fiber but are inadequate for a beating fiber. First, the clearly recognized Ca entry via slow inward channels comprises a substantial amount of the [Ca] change involved in the initiation of contraction. The removal of this Ca in the steady state must be via Na/Ca exchange, which constitutes the principal Na load for a beating heart. Second, Na/Ca exchange will be reversed during the plateau of the action potential; thus Ca will enter and Na will be pumped out by this mechanism. To the extent that Ca entry via channels and carrier are equal in magnitude, the Na pumped out avoids a future load on the Na/K mechanism. Changes in $[Na]_i$ brought about by partial inhibition of the Na pump will affect the steady-state level of $[Ca]_i$ and the kinetics with which the Na/Ca carrier moves Na both inward and outward. A greater fraction of Na extrusion will be by Na/Ca exchange; stated differently, more Ca entry by Na/Ca exchange during depolarization will occur in ouabain-treated fibers.

It is instructive to make certain changes in this coupled [(Na/K) − (Na/Ca)] system and observe its behavior. Removal of Na_o leads to a decrease of Ca efflux, an increase in Ca influx, and a transient hyperpolarization as the Na/K system continues to produce an outward current while the inward current of Na/Ca is replaced by an outward current (Na emerging in exchange for Ca). In the steady state, Na will have been pumped out; there will be neither a hyperpolarization nor a depolarization (i.e., an E_m based solely on diffusion will exist). The leak of Ca in and the compensating leak of K out will continue, although both pumps (Na/Ca and Na/K) will have ceased functioning.

If Ca_o rather than Na_o is removed, Ca_i is pumped down to low levels. There then is no further Na entry, while the Na pump reduces $[Na]_i$ to low levels. Both pump currents effectively vanish. This emphasizes the high degree of coupling that exists between Na/K and Na/Ca pumping, since both require a common ion for transport.

A somewhat different approach is to selectively inhibit the Na/K pump with, for example, ouabain. Since Na efflux via the pump is less than Na influx via Na/Ca exchange, $[Na]_i$ rises. Thus there is an enhanced Ca influx and an inhibited Ca efflux via Na/Ca exchange; hence $[Ca]_i$ rises. New levels of both $[Na]_i$ and $[Ca]_i$ occur, at which a balance between Na entry via Na/Ca exchange and Na pumping can be reached. With total inhibition of Na/K exchange, $[Na]_i$ rises to a level where the quantity of Na entering in exchange for Ca equals the quantity of Na leaving in exchange for Ca entering. A steady state with respect to $[Na]_i$ now depends on the presence of Ca_o rather than on K_o. From an energetics point of view, the cell must have reached an equilibrium, either by changes in E_m (so that $E_m = E_R$) or by changes in E_{Na} and E_{Ca} (such that the Na/Ca exchange mechanism is in equilibrium).

As a practical matter, the cardiac cell, as a result of its long-duration depolarization during an action potential, radically changes its electrochemical gradient for Na as a result of the membrane potential change from the resting level to the plateau. This change results in a flow of Ca that is now inward rather than outward. Modulation of the magnitude of such flows in both directions can take place if poisons, such as ouabain, are employed to partially inhibit the Na/K pump. The resultant change in $[Na]_i$ alters E_{Na} and hence the equilibrium point of Na/Ca exchange. Equilibrium requires $[Ca]_o([Na]_i)^4K = [Ca]_i([Na]_o)^4(1/K)$, where K is a term in membrane potential. If K, $[Ca]_o$, and $[Na]_o$ are held constant, then a 10% increase in $[Na]_i$ produced by ouabain requires a 46% increase in $[Ca]_i$. Therefore, there is (a) a higher resting $[Ca]_i$, and (b) a greater Ca influx upon depolarization since influx depends on $[Ca]_o([Na]_i)^4$.

If $[Ca]_o$ is varied while $[Na]_i$ is measured via an ion-specific microelectrode (1), then a 10-fold increase in $[Ca]_o$ decreases $[Na]_i$ to about one-half, but there is also a membrane potential change that was not controlled. The experiment does show that the exponent of $[Na]_i$ must be greater than three to hold $[Ca]_o([Na]_i)^r$ constant.

Na EFFLUX VERSUS $[Na]_i$

Measurements of Na efflux in RBC at a $[K]_o$ of about 2 mM show that this is half-maximal at a $[Na]_i$ of about 30 mM (2). Measurements by DiPolo (3) show that the Na-dependent Ca influx in squid axons in half-maximal at a $[Na]_i$ of about 60 mM. Since Ca influx and Na efflux are coupled, $[Ca]_o$-dependent Na efflux is half-activated at this concentration. In a fiber with both a Na/K and a Na/Ca pumping system operating, the expected result of this dual Na extrusion is that the efflux from the two pumps will sum. The expected result is similar to that shown in Fig. 6.1, in which a Na/K pump curve and a $[Ca]_o$-dependent efflux of Na versus $[Na]_i$ are shown separately, along with the sum of these curves. The dashed line is a reasonable approximation to a straight line, an arrangement found, for example, in squid giant axons for Na efflux versus $[Na]_i$.

Corresponding measurements on mammalian tissues of the [Na] for half-activation of the Na/Ca system have not been made. Indi-

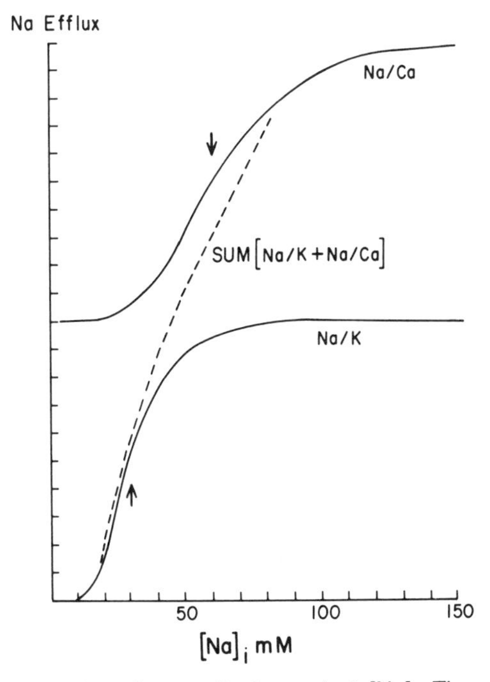

FIG. 6.1. Na efflux plotted as ordinate against $[Na]_i$. The curve labeled Na/K is the activation of the Na/K pump with a $K_{1/2}$ of 30 mM $[Na]_i$. The curve labeled Na/Ca is the efflux of Na produced by the Na/Ca exchange running backward with a $K_{1/2}$ of 60 mM. *Dashed line,* sum of these two Na efflux-producing mechanisms.

rect measurements, however (discussed below), suggest that it may not differ greatly from 60 mM. A feature of these curves is that with normal values for $[Na]_i$ in most animal cells of between 15 and 30 mM, the Na pump is working at the midrange of its capabilities, while the Na/Ca system is working in a range where Ca-dependent Na efflux is a power function of $[Na]_i$. This means that a small change in $[Na]_i$ can catalyze a large Ca entry, or precisely the kind of result that is observed by the application of clinical dose levels of cardiac glycosides.

A second feature of the curves is that if Na/K pumping were abolished, then Na regulation of the fiber could take place by the operation of Na/Ca exchange, provided that $[Na]_i$ were to rise to levels that could develop the appropriate level of Na efflux to balance Na inward leak. This prediction is in good agreement with the results of Deitmer and Ellis (1), who showed that in quiescent Purkinje fibers, $[Na]_i$ was of the order of 7 mM in normal Ringer; upon maximal ouabain poisoning, however, $[Na]_i$ rose to about 25 mM, at which level it stabilized. It was also clear that fibers could gain Na in Ca-free solutions and then, by the application of Ca_o-Ringer's, $[Na]_i$ could be reduced to lower levels. The results of Deitmer and Ellis (1) are persuasive to the view that cardiac fibers have two independent systems capable of controlling $[Na]_i$.

A conclusion from Deitmer and Ellis (1) is that Ca_o is also capable of reducing $[Na]_i$, and that this effect is independent of the presence or absence of ouabain. This information supports the notion that there are two separate pathways in Purkinje fibers designed to produce a Na efflux. The existence of such pathways also lends support to the idea already advanced that there are multiple sources of Na influx to provide the ion that must be pumped out. In addition to the conventional Na entry via Na channels during the rapidly rising phase of the action potential, conditions connected with the Na/Ca exchange mechanism itself also provide Na entry.

One of the difficulties with accepting the notion that the action of ouabain and other related drugs on heart arose solely from the action of such substances on Na/K transport was that the change in $[Na]_i$ was so small. The change produced by clinical doses of ouabain is of the order of 10 to 12% (4), an alteration too small to measure in cardiac tissue, given the variation of extracellular space corrections. This changed $[Na]_i$ produces a similar percentage change in Ca_i. From the foregoing discussion, however, it is apparent that a larger change in $[Ca]_i$ has been produced during the systolic interval. Because ionized [Ca] is related to $[Na]_i$ as a fourth power function, during depolarization a significant fraction of the Ca entering the fiber comes from Na/Ca exchange. The analytically measured increase in Ca_i is stored Ca rather than $[Ca]_i$. Hence increased mechanical response in ouabain is assumed to be a result of more Ca released from stores during depolarization, as well as more Ca entry from Na/Ca exchange.

MEMBRANE POTENTIAL OF CARDIAC FIBERS

The existence of a membrane potential was recognized in electrophysiology even before the technical ability existed for its measurement. On the simplest basis, this potential was a K diffusion potential and could be expressed as a Nernst potential for K. As actual measurements became possible, it was clear that the potential was less than E_K. A shunt on this theoretical limit was considered to be a nonideality in the inward leakage of cations, such as Na. The situation resembled the difference between the potential of a battery with and without current flow. If the shunt allowed sufficient current flow, there was an *IR* drop (where *I* is current and *R*, resistance) which could be described as

$$I_K + I_{Na} = 0$$

$$I_K = g_K(E_m - E_K), I_{Na} = g_{Na}(E_m - E_{Na})$$

where g is conductance, or $1/R$.

$$E_m = \frac{(g_K E_K + g_{Na} E_{Na})}{g_K + g_{Na}}$$

Since $g_K/(g_K + g_{Na}) = T_K$, the transference number or fraction of current carried by the ion, this expression is $E_m = (T_K E_K) + (T_{Na} E_{Na})$. This arrangement was satisfactory only if the mechanism that restored the Na gradient was electroneutral and moved equal quantities of Na and K in opposite directions across the membrane. Experiment quickly showed that charge movement via the Na/K pump was not equal, and that more Na^+ were extruded than K^+ taken up. This finding meant that the Na pump would produce a metabolically dependent contribution to the membrane potential that could be described roughly as

$$E_m = E_K - (j\mathrm{Na}_o + k\mathrm{Na}_i)R_m$$

where $j\mathrm{Na}_o$ is the passive net Na flux that is producing an inward current, $k\mathrm{Na}_i$ is the contribution of the Na pump to the outward current, and R_m is membrane resistance. In the steady state, a simple arrangement would be for the Na pump to balance Na leak and without coupling the Na extrusion to any other ion movement. Then, since $(j\mathrm{Na}_o + k\mathrm{Na}_i)$ would sum to zero in the steady state (by definition, Na influx and efflux are equal), the membrane potential would be simply E_K.

All actual Na pumps appear to be Na/K pumps with a Na flux larger than the K flux (i.e., a coupling ratio Na/K > 1). This arrangement means that again for the steady state, E_m will be less than E_K, since some K will be introduced into the fiber when Na fluxes are in balance. This clearly is a shunt against the limiting potential E_K. A stable membrane potential means that ion currents generated by either diffusion or metabolic or gradient-dependent processes must sum to zero. Thus

$$I_{\mathrm{Na}} + I_{\mathrm{K}} + I^{\mathrm{P}}_{\mathrm{Na}} + I^{\mathrm{P}}_{\mathrm{K}} = 0$$

where I^{P} is a symbol for the pumped current generated by the Na/K pump. The coupling ratio of the pump $r' = I^{\mathrm{P}}_{\mathrm{Na}}/I^{\mathrm{P}}_{\mathrm{K}}$. Thus $I^{\mathrm{P}}_{\mathrm{Na}} = r'I^{\mathrm{P}}_{\mathrm{K}}$, and a steady state requires that

$$I_{\mathrm{Na}} = \mathrm{I}^{\mathrm{P}}_{\mathrm{Na}}, \qquad I^{\mathrm{P}}_{\mathrm{K}} = I_{\mathrm{K}}$$

Hence

$$I_{\mathrm{Na}} = r'I_{\mathrm{K}}$$

Expressions for I_{Na} and I_K are

$$I_{Na} = P_{Na}([Na]_o - [Na]_i \exp(-E_m F/RT)) f(E)$$

$$I_K = P_K([K]_o - [K]_i \exp(-E_m F/RT)) f(E)$$

where $f(E)$ is some unspecified potential dependence of the Na and K currents. Rearranging yields

$$E_m = \frac{P_{Na}[Na]_o + r'P_K[K]_o}{P_{Na}[Na]_i + r'P_K[K]_i}$$

The following are properties of this equation: If r' is large, the expression reaches that for a Nernst potential for K; if the value of r' is unity, the expression is identical with the Goldman-Hodgkin-Katz equation. E_m can never be greater than E_K in the steady state. While some experimental treatments can move the membrane potential transiently to values greater than E_K, the effect of this change will be to start a change in E_K that will move it or the membrane potential into compliance with the equation given above.

Experimenters often substitute Na_o with some other cation, such as Li. A description of the membrane potential now becomes more complicated because: (a) a reduction in $[Na]_o$ allows $[K]_o$ to become more effective in stimulating Na/K exchange; (b) often the cation replacing Na_o is not capable of being extruded from the cell; hence no steady state with respect to $[K]_i$ is possible; and (c) the decrease in $[Na]_i$ as a result of lowering $[Na]_o$ can be expected to substantially interfere with Ca entry during depolarization, since entry is proportional to $[Ca]_o([Na]_i)^4$.

Qualitatively, the instantaneous effect of a reduction in $[Na]_o$ can be expected to be hyperpolarization if: (a) the substitute cation is less permeable than Na; (b) the Na pump is stimulated more effectively by $[K]_o$; and (c) more Ca entry in exchange for Na_i takes place, thus resulting in a net Na exit (hyperpolarization).

When $[Na]_i$ assumes a new time-independent value, there is less of a contribution from (b) and (c) above, so that E_m becomes much closer to a diffusion potential specified by E_K less a shunt for entering cation. Note that (a) is seldom a factor in the membrane potential equations, since substitutes for Na (Li, choline) often have the same permeability as Na.

The foregoing emphasizes that additional electrogenic mechanisms in the membrane will contribute to the membrane potential and must be included. In cardiac cells in particular, a term in Ca pumping seems to be required, since the entry and removal of Ca are effected by a Na countertransport. A first approach is to consider the fiber similar to a resting nerve fiber in which the Na pump removes Na previously admitted during nervous activity. In the case of Ca, the entry of this ion during bioelectric activity is then reversed by Na/Ca exchange whereby nNa enter per Ca pumped out. The evidence is that n is greater than 2, thus making the operation of the countertransport mechanism rheogenic. When $n = 4$, there are two net charges moving inward per Ca expelled. This movement can be expected to exert a depolarizing influence on the membrane potential. Note that in the steady state, this entering Na will be extruded by the Na/K pump and will contribute to hyperpolarizing the membrane, as described above.

The following treatment is from Sjodin (5). For Ca^{2+} subjected to a constant electrical field across the membrane, the general equation for ionic flux is

$$I_{Ca} = P_{Ca}\left(\frac{zE_mF}{RT}\right)\left(\frac{[Ca]_o - [Ca]_i \exp(zE_m F/RT)}{1 - \exp(zE_m F/RT)}\right) \quad [1]$$

where I_{Ca} is current, P is permeability coefficient, z is ionic valence, and E_m is membrane potential. All other symbols have their usual significance. The first equation sought is obtained by solving the sum $2I_{Ca} + I_K + I_{Na} + I_{Cl} = 0$. The expression for E_m with Ca permeability taken into account is

$$E_m = \frac{RT}{F}\ln\frac{4P'_{Ca}[Ca]_o + P_K[K]_o + P_{Na}[Na]_o + P_{Cl}[Cl]_i}{4P''_{Ca}[Ca]_i + P_K[K]_i + P_{Na}[Na]_i + P_{Cl}[Cl]_o} \quad [2]$$

$$P'_{Ca} = P_{Ca}\frac{\exp(E_m F/RT) - 1}{\exp(2E_m F/RT) - 1} \quad [3]$$

and

$$P''_{Ca} = P'_{Ca}\exp(E_m F/RT) \quad [4]$$

Given the form of equation [1] for divalent cations, it is not possible to obtain a solution having the form of equation [2] without defining potential-dependent "apparent" permeability coefficients P'_{Ca} and P''_{Ca}. The transcendental nature of equation [2] means that solutions must be made by simple trial and error or by computer, using equations [3] and [4]. For normal values of the resting potential, however, $P'_{Ca} \cong P_{Ca}$. It can be verified easily that equation [2] reduces to the equilibrium potential for Ca ions for a perfectly Ca-selective membrane.

$$E_{Ca} = (RT/2F) \ln ([Ca]_o/[Ca]_i) \qquad [5]$$

With

I^P_{Na}, I^P_K as the coupled Na/K pumped currents

J^P_{Na}, J^P_{Ca} as the currents produced by Na/Ca transport

we can, for the steady state, write

$$-I_{Ca} = J^P_{Ca} \qquad [6]$$

$$I_{Na} + J^P_{Na} = I^P_{Na} \qquad [7]$$

$$-I_K = I^P_K \qquad [8]$$

The Na/K pump coupling ratio, r', has previously been defined (6).

$$r' = \frac{m_{Na}}{m_K} \qquad [9]$$

The stoichiometric factor, r, for Na/Ca transport is the number of Na ions transported per Ca ion:

$$J^P_{Na} = -rJ^P_{Ca} \qquad [10]$$

To obtain the required steady-state solution, the sum

$$I_{Na} + I_K + 2I_{Ca} + I^P_K + I^P_{Na} + 2J^P_{Ca} + J^P_{Na}$$

is set equal to zero to fulfill electrical neutrality. Substitutions are made from equation [1] and from equations [6] to [10]. The resulting equation for E_m is

$$E_m = \frac{RT}{F} \ln \frac{2rP'_{Ca}[Ca]_o + r'P_K[K]_o + P_{Na}[Na]_o}{2rP''_{Ca}[Ca]_i + r'P_K[K]_i + P_{Na}[Na]_i} \qquad [11]$$

It is of interest to note some features of equation [11]. If the stoichiometric ratio $r = 2$, two Na ions are transported per Ca ion, and Na/Ca transport is electrically neutral. In this case, equation [11] reduces to equation [2], where Ca ions exert only their passive electrodiffusional effect. The actual value for r appears to be 4 (ref. 7). In this case, the terms due to Ca are twice those due to the electrodiffusional effect of Ca ions. For the values of $[\mathrm{Ca}]_i$ in resting excitable cells, the effect of Na/Ca transport is an additional depolarization numerically equivalent to that obtained by doubling the value of either P_{Ca} or $[\mathrm{Ca}]_o$. The source of the depolarization is the inward transport of two charges during each cycle of Na/Ca transport operation. The precise equivalency of the electrical effects of I_{Ca} and Na/Ca transport is attributable to the fact that $J^{\mathrm{P}}_{\mathrm{Ca}}$ just balances I_{Ca} and that net charge on the Na/Ca carrier happens to equal 2, the valence of Ca ions. Similar interpretations can be made for other values of r. Given the very low values for $[\mathrm{Ca}]_i$ in most excitable cells, the Ca term on the bottom of equation [11] can be ignored.

When ionic concentrations are permitted to change, equations [6] to [8] must be modified. The ratio of transport rate to leakage rate, f, can be defined for Ca ions by the relationship:

$$f_{\mathrm{Ca}} = J^{\mathrm{P}}_{\mathrm{Ca}}/I_{\mathrm{Ca}} \qquad [12]$$

Similarly, for K ions,

$$f_{\mathrm{K}} = I^{\mathrm{P}}_{\mathrm{K}}/I_{\mathrm{K}} \qquad [13]$$

For Na ions, the factor f_{Na} must also include the inward movement due to Na/Ca transport:

$$f_{\mathrm{Na}} = I^{\mathrm{P}}_{\mathrm{Na}}/(J^{\mathrm{P}}_{\mathrm{Na}} + I_{\mathrm{Na}}) \qquad [14]$$

Applying the same condition for electrical neutrality applied previously and substituting from equations [9], [10], and [12] to [14] yields

$$rf_{\mathrm{Ca}}I_{\mathrm{Ca}} + (f_{\mathrm{K}}/f_{\mathrm{Na}})\, r'I_{\mathrm{K}} + I_{\mathrm{Na}} = 0 \qquad [15]$$

Equation [15] is the algebraic summary statement of the condition for electrical neutrality. This equation clearly places restrictions on the values that f_{Ca}, f_K, and f_{Na} can assume. If, for example, one is in the steady state and the Ca transport rate is increased such that f_{Ca} assumes a value different from unity, the remaining fluxes must become adjusted in such a way that equation [15] is obeyed. Making substitutions for I values from equation [1] and solving for E_m, we have

$$E_m = \frac{RT}{F} \ln \frac{2rf_{Ca}P'_{Ca}[Ca]_o + r\,(f_K/f_{Na}\,P_K[K]_o + P_{Na}[Na]_o}{2rf_{Ca}P''_{Ca}[Ca]_i + r\,(f_K/f_{Na})\,P_K[K]_i + P_{Na}[Na]_i} \qquad [16]$$

which gives the membrane potential for nonsteady-state conditions. If the coupling ratios for the two transport processes and the magnitudes of all the flux components are known, the value of the membrane potential can be computed, provided ionic permeability coefficients and concentrations are known. The equation for the steady state, [11], can be regarded as a special case of equation [16], for which $f_{Ca} = f_K = f_{Na} = 1$.

Increasing the value of f_{Ca} above 1.0 (meaning that more Ca ions are transported outwardly than leak inwardly), results from equation [16] in increasing depolarization since the increased Ca transport must be accompanied by an increased inward movement of charge via the Na/Ca carrier.

The change in membrane potential resulting from a change in the Ca transport rate via the Na/Ca carrier depends on the nature of the flux readjustments. The presence of three mechanisms for ion movement makes possible the achievement of electrical neutrality. To obtain useful solutions to equation [16], careful attention must be paid to the physiological events occurring during a readjustment of the ionic fluxes.

To illustrate the use of equation [16], the membrane potential can be plotted as a function of increasing Na/Ca transport rate for different permeability coefficient ratios. This is done in Fig. 6.2, in which permeability coefficients and ionic concentrations remain constant for each curve. The potential plotted is the instantaneous value assumed when the indicated Ca transport rate occurs before ionic concentrations have had a chance to change. The value of the

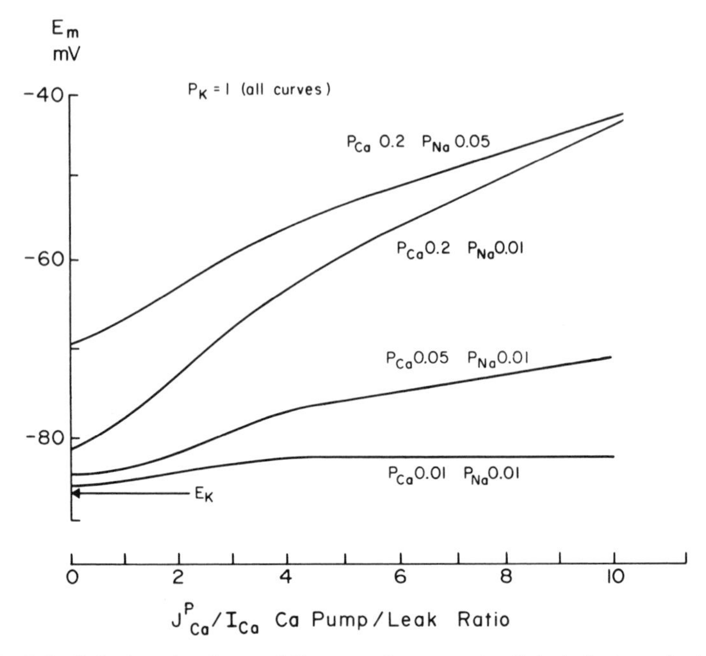

FIG. 6.2. Calculated values of the membrane potential plotted against increasing Ca transport rate, t_{Ca}, expressed as the fraction J^{P}_{Ca}/I'_{Ca}, where I'_{Ca} is the passive inward Ca^{2+} leak in the steady state (*arrow*). At $J^{P}_{Ca} = 0$, equation [2] was used; in the steady state, equation [11] was used. All remaining values were calculated from equation [16], with the aid of equation [15]. Ion electrodiffusion rates were calculated using equation [1], with the following ionic concentrations (in mM): $[Na]_o = 150$; $[K]_o = 5$; $[Ca]_o = 2$; $[Na]_i = 10$; $[K]_i = 150$. The normal value of $[Ca]_i$ is about 0.1 μM or less, making the Ca term in the denominator negligible in all cases. The value of p_K was normalized to 1.00 for all calculations. Curves are plotted for various representative permeability coefficient ratios. The value used for the Na/Ca transport stoichiometric ratio, *r*, was 4.0 for all calculations.

potential for zero Na/Ca transport rate is simply the appropriate solution of equation [2]. The system is not in a steady state at this point. At the point $J^{P}_{Ca}/I_{Ca} = 1$, the system was chosen to be in the steady state for Na, Ca, and K ions. The potential is the solution of equation [11]. The steady state for Ca ions for all permeability ratios plotted is achieved at the cost of only a few millivolts depolarization. As ionic concentrations are not permitted to change for these curves, the Na/K pumping rate remains constant, and the increased inward charge movement due to the Na/Ca carrier must be balanced by ionic electrodiffusional leaks.

If we consider the passage of time, however, the value of $[Na]_i$ must increase as a net Na influx must occur when $J^P_{Ca}/I_{Ca} > 1$ under the assumed conditions. As $[Na]_i$ rises, the Na/K pump rate rises if the pump has not reached the saturation rate. Under these conditions, the hyperpolarizing effect of the Na pump becomes apparent by causing some of the increased Na/Ca transport charge to be balanced by Na outward pumping rather than by ionic leaks.

These changes of potential are illustrated in Fig. 6.3 for P_K:P_{Ca}:P_{Na} = 1:0.2:0.01. At the point marked by the arrow, the steady state is perturbed by increasing the J^P_{Ca}/I_{Ca} ratio from 1 to

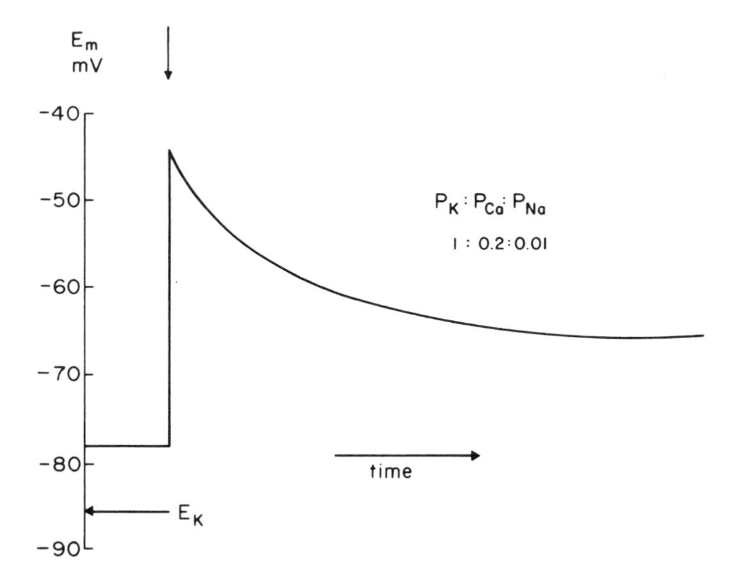

FIG. 6.3. Calculated resting membrane potential plotted versus time in arbitrary units when the steady state is perturbed by a sudden 10-fold increase in the Ca outward transport rate. The stoichiometry of Na/Ca transport demands an accompanying 10-fold increase in the Na inward transport rate by the Na/Ca carrier. (See text for explanation of the curve.)

10. Such a change might be produced, for example, by an internal release of Ca that greatly increased $[Ca]_i$, or it might be produced by a large Ca entry. At any rate, the change would appear as an inward (depolarizing) current initially accompanied by a substantial increase in $[Na]_i$. As $[Na]_i$ rises, the increased level of Na/K pump operation produces hyperpolarization. The value of $[Na]_i$ can only rise, however, until the Na pump balances the Na leakage in, plus the new value of the Na influx by Na/Ca transport. At this point, Na ions are in a new steady state, and further hyperpolarization cannot occur. Not all the increased inward charge movement by Na/Ca transport has been compensated by electrodiffusion fluxes, mainly that due to K^+. The final potential reached is thus a steady state for Na but not for K and Ca, whose net fluxes balance at the final potential reached.

Another feature of equation [16] is that the value of f_{Ca} need not be positive. As the value of I_{Ca} is such as to always be in the inward direction for values of E more negative than E_{Ca}, f_{Ca} becomes negative if the direction of Na/Ca transport is reversed when E is in this range (equation [12]). It is known that Na/Ca transport can operate in the reversed mode under certain conditions (8). Under these conditions, Na ions are transported outwardly over the Na/Ca carrier, and Ca ions are transported inwardly. According to equation [16], the reversed direction of Na/Ca transport gives rise to a hyperpolarizing effect, as f_{Ca} is now negative and the Ca term in the numerator subtracts numerically from the other terms. The source of the hyperpolarization is the outward movement of charge by the Na/Ca carrier. Readjustments of the Na/K pump must be taken into account here as well.

ELECTROGENIC Na TRANSPORT IN PURKINJE FIBERS

It has been clear for some years that Purkinje (and presumably other cardiac) fibers have an electrogenic transport mechanism for Na that resembles in most respects that found in other tissues (9–11). In particular, the 1975 study by Isenberg and Trautwein (11) showed that voltage clamp current (2 sec after the start of a pulse) was reduced rather uniformly by cooling over a range of potentials from −80 to −40 mV, as shown in Fig. 6.4. A similar result was obtained using dihydro-ouabain. A most significant finding was that with dihydro-ouabain and cooling used together, the current-voltage relationship was the same as cooling alone in the range of potentials from −80 to −40 mV but had a greatly reduced slope both negative to −80 and positive to −40. This finding suggests that other carrier-mediated processes that are not the Na/K pump are contributing to the current-voltage relationships in these regions of membrane potential. It is also worth noting that membrane current in the steady state is virtually zero in ouabain-treated or cooled fibers for the potential range −80 to −40 mV.

A new technique for studying electrogenic Na pumping in Purkinje fibers has been developed by Gadsby and Cranefield (12). It involves momentary loading of the fibers with Na, followed by the unloading of such fibers by Na pumping. The entire procedure is carried out under voltage clamp conditions (at the lower resting

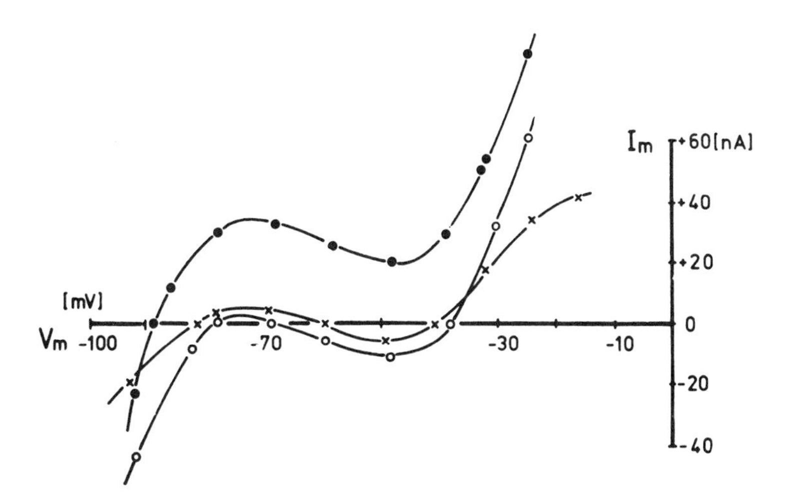

FIG. 6.4. Plot of membrane current versus voltage for Purkinje fibers held at the potential indicated for 2 sec before the measurement was made. *Solid circles,* normal fibers; *open circles,* cooling; *crosses,* cooling and dihydro-ouabain used together. (Data taken from ref. 11.)

potential of the fiber or about −40 mV). A record of such an experiment where K-free Ringer's is used to slow the Na pump so that the fiber will load with Na is shown in Fig. 6.5. Under K-free conditions, there is an inward current. Upon switching to 4-mM K Ringer's solution, this changes to an outward current that quickly declines to zero. The same experiment, repeated in the presence of acetyl strophanthidin, showed no trace of an outward current, while removal of this drug allowed recovery of the current. The evidence here is compelling that the current measured as a transient in the voltage clamp is that produced by the operation of the Na pump.

A second feature of these measurements is that the inward current that develops when K-free conditions are imposed becomes less with time and extrapolates to zero after 6 min in K-free solution (see Fig. 6.5). This implies that as $[Na]_i$ increases, this inward current decreases. Several explanations are possible. Perhaps the simplest is that the inward current is I_{Na}, and that as $[Na]_i$ rises, a point is reached where $E - E_{Na} = 0$. This is impossible, since E_m is clamped to −40 and an E_{Na} of −40 would lead to severe osmotic imbalance. A second possibility is that increases in $[Na]_i$ lead to increases in outward K current. Finally, increases in $[Na]_i$ may lead

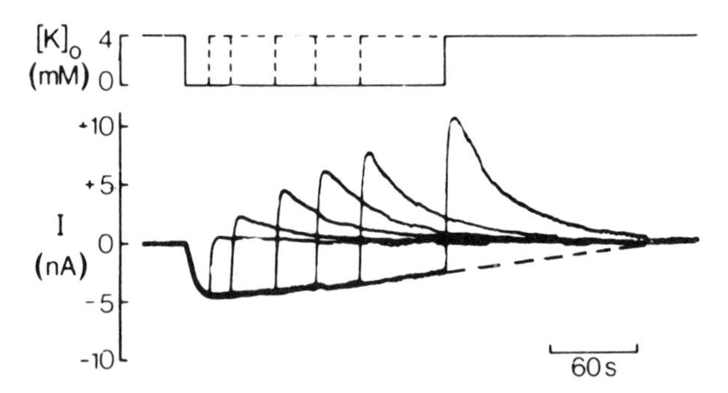

FIG. 6.5. Recording of membrane current as a function of time when the bathing membrane around Purkinje fibers was changed from 4 to 0 mM K (*top tracing*). Note the short 15-sec statement of the preparation with K-free solution against an inward current but no outward current upon return to normal K medium; longer periods of immersion produced progressively larger electrogenic currents from the Na pump. *Dashed line,* extrapolates the inward current to zero. (From ref. 12.)

to the activation of Na/Ca exchange in the "reverse" direction, i.e., where Ca entry promotes Na efflux and an associated outward current.

Ellis (13) and Deitmer and Ellis (1) also studied the Na/K pump in Purkinje fibers. The method of measurement of $[Na]_i$ was by a Na-sensitive microelectrode. Although problems do exist with ion-sensing electrodes in cardiac cells, *changes* in $[Na]_i$ probably are adequately sensed by such devices. The measurements show that upon application of K-free Ringer's, $[Na]_i$ rises, as it does when strophanthidin is applied. In both cases, it reaches a value far from the equilibrium value that would be expected if there were no Na extrusion and Na entry continued at its normal rate. Because the Deitmer and Ellis (1) measurements were not made under voltage clamp of the membrane, the membrane potential varied substantially with K-free conditions but did not vary so much when strophanthidin was applied. In both cases, however, a stable $[Na]_i$ was reached after a few minutes, which was shown to depend on Ca_o. Whereas the Na efflux that depends on the Na/K pump has been abolished by a cardiac glycoside, a new Na efflux dependent on $[Ca]_o$ has taken over the regulation of $[Na]_i$ at a value substantially higher than normal but less than equilibrium.

SUMMARY

In summary, Isenberg and Trautwein (10, 11) have shown that in the absence of Na pumping, virtually no membrane current is produced by Purkinje fibers over the potential range -80 to -40. The Na/K pump produces a constant displacement of the current voltage relationship. Gadsby and Cranefield (12) have shown that at lower resting potential of Purkinje fibers (-40), the magnitude of the pump current is enhanced by temporarily increasing $[Na]_i$. Deitmer and Ellis (1) showed that a stable upper level of $[Na]_i$ exists in Purkinje fibers in the presence of strophanthidin which is maintained by Ca_o.

The apparent constancy of the Na pump current must be reconciled at -80 and -40 mV. The pump must receive the Na that it pumps out, yet it is also reasonable to suppose that I_{Na}, the steady background current of Na flowing through channels, is larger at −40 than at −80 mV. A carrier current is expected that involves both Na and Ca; it must move Na inward when E_m is more negative than E_R, the carrier reversal potential. Finally, the Na pump current should be potential independent. These various currents are diagramed in Fig. 6.6. Note that at −80 mV, virtually all of Na being pumped out is coming from the Na/Ca exchange mechanism; hence internal stores of Ca are being depleted. At −40, the assumed reversal potential of the Na/Ca exchange, all the Na for pumping is entering via the Na channels. An interesting feature of this relationship, not yet examined experimentally, is that at a membrane potential of zero, the I_{Na} pump current and I_C, the carrier current, sum to produce a Na efflux twice as large as that of the Na/K pump, while I_{Na} may also be larger. All these curves are drawn for the instantaneous case; imbalance in fluxes will lead to changes in ionic concentrations of Ca even though Na fluxes are in balance. The

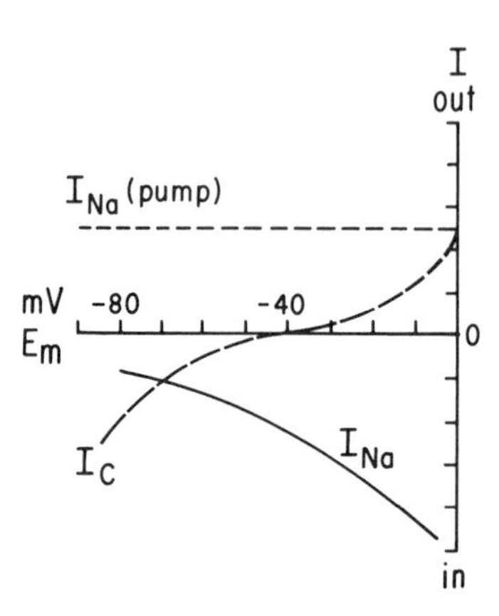

FIG. 6.6. Current-voltage plot for the current generated by the Na pump on the assumption that there is a constant Na load, which comes from various sources. The line labeled I_C is the assumed carrier current and is by Na/Ca exchange and with a reversal potential at −40. The solid line labeled I_{Na} is the way that a state of inward leak of Na is assumed to vary with membrane potential. Note that there are three points where the Na fluxes sum to zero (i.e., $[Na]_i$ is in a steady state): these points are −80, −40, and 0.

diagram suggests that known points of membrane potential stability (−80, −40) may be the result of a balance of Na fluxes. An additional point of apparent membrane potential stability (the plateau, near zero E_m) could also be stabilized by a balance of Na fluxes.

Ca IN INTERNAL COMPARTMENTS

Introduction

In addition to the other complications in studies of Ca movement in cardiac cells, it has long been recognized that to an extent that depends on a specific cardiac cell type, greater or lesser amounts of Ca are stored in internal compartments, and some of this Ca is releasable during excitation.

In the case of all muscle contraction, it is generally conceded that $[Ca]_i$ must change from a low value appropriate to relaxation (<50 nM) to a much higher value (>1 μM). In skeletal muscle, this change in $[Ca]_i$ is substantially independent of $[Ca]_o$; in cardiac muscle, mechanical activity ceases upon removal of Ca_o, although action potential generation continues. Upon restoration of Ca_o, mechanical activity returns with each of several contractions increasing in strength until a steady state is reached. It has been argued that since I_{si} does not provide enough Ca for contraction, there must be a Ca-induced Ca release responsible for this and other effects. This controversy is not discussed here; what is important to recognize is that in some tissues (frog atria), it is likely that all the Ca for contraction comes from outside the fiber (14), while for other cardiac tissues, there may be substantial internal release.

The Sarcoplasmic Reticulum

Analytic values for Ca in cells range from 50 to 1000 μmoles/kg. Recent studies emphasize that all but a minute fraction of this Ca is bound, complexed, or stored in internal compartments. In the case of the squid giant axon, it has been possible not only to measure free ionized Ca but to show that at least two compartments store Ca: one is FCCP or CN sensitive and is presumably the mitochondria, while the other is likely to be the smooth endoplasmic reticulum.

In the case of muscle, it has long been recognized that a free [Ca] below about 100 nM is necessary for relaxation and that there exists in muscle a membranous structure, the sarcoplasmic reticulum (SR), that is the principal storage site for Ca. Also, mitochondria as well as chemical compounds are capable of complexing Ca.

Preparative biochemical methods have been developed for obtaining both intact mitochondria and SR vesicles. Both preparations show Ca-accumulating ability *in vitro*. Studies with such preparations make it possible to specify that mitochondria are capable of accumulating Ca inside to a concentration of 10^3 that of the ambient; SR vesicles can accumulate free Ca to 10^4 that of the outside concentration. The diagram in Fig. 6.7 shows the relationships between these Ca-storing entities and the [Ca] ratio across the sarcoplasmic membrane discussed earlier, which is 10^5.

Both the SR membrane and the sarcoplasmic membrane have energy-requiring ion pumps for Ca and separate gating mechanisms that allow Ca to move by diffusion. Gating mechanisms for the SR Ca release are unknown, while the SR Ca pump has been studied intensively. The assumption is made here that an electrical signal

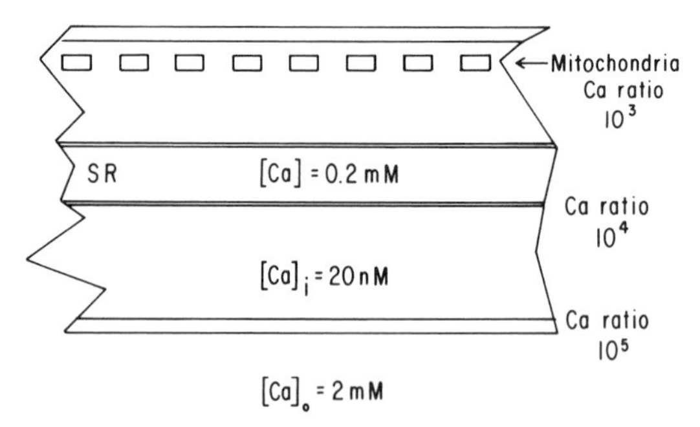

FIG. 6.7. Diagram showing the distribution of Ca_o, Ca_i, Ca_{SR}, and Ca in the mitochondria. The accumulation ratios for the Ca in SR and mitochondria are those obtained by *in vitro* studies of isolated preparations.

in the sarcoplasmic membrane is translated into a gating of Ca from the SR store. The Ca gating must inactivate, since a maintained depolarization of the surface membrane leads to a spontaneous relaxation. This implies that the SR is able to pump down the high Ca of the fiber to levels below the threshold for mechanical activation.

Cells capable of contraction differ greatly in the extent to which they rely on a SR to provide Ca for their mechanical response. In the case of fast twitch skeletal fibers, the reliance is total; in the case of certain frog cardiac fibers, the SR is virtually nonfunctional. Although mammalian cardiac fibers are somewhere between these two extremes, no cardiac fibers are as independent of $[Ca]_o$ as is a skeletal fiber. Thus for every contraction, a substantial fraction of the total Ca required must come from outside the fiber. Stated otherwise, the SR is capable of substantial modulation of both contraction and relaxation.

In addition to recognizing the SR in cardiac muscle, it is necessary to describe the compartment structure that connects Ca fluxes across the sarcolemma with those into the SR. Several arrangements are possible (Fig. 6.8): a series and a parallel connection. Even more elaborate arrangements have been proposed, whereby Ca entering via the surface membrane goes into a store that in turn feeds a second store in series with the first one; this is termed the ''elaborate'' arrangement. Purely anatomical considerations favor the parallel arrangement, whereby Ca can be fed from the surface membrane or from the SR to the contractile machinery.

Other arrangements have been proposed because physiological measurements show that a supposed increased Ca entry is not immediately reflected in increased tension during an action potential. This point is examined later; here it is sufficient to note that (a) Ca entry is not unambiguously determined by measuring slow inward current; Ca entry via Na/Ca exchange has not been considered, and (b) until the present, little attention was given to the role that $[Na]_i$ might play in determining the amount of Ca entry during a depolarization or, indeed, the amount of Ca stored in the SR.

It is sometimes assumed that Na/Ca exchange, operating in a 2

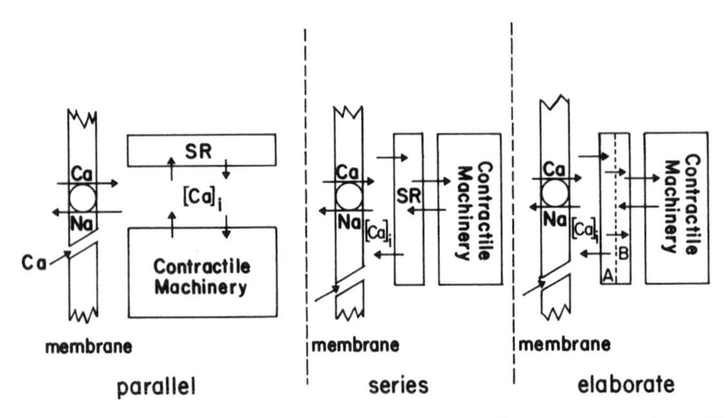

FIG. 6.8. Possible compartment structures connecting the extracellular fluid with the SR are shown for a "parallel," "series," or "elaborate" arrangement. In the parallel case, Ca entry by channel or by carrier is into the myoplasm. Both the SR and the contractile machinery have access to this Ca. In the series arrangement, Ca entering must transit the SR before it can interact with the contractile machinery. This arrangement requires a high degree of electrical coupling between the surface membrane and the SR; it is difficult to see why Na-free solutions produce contraction on the basis of this arrangement. Finally, the elaborate arrangement requires that Ca in one compartment of the SR move to a second compartment before it becomes available for interaction with the contractile machinery.

Na:1 Ca stoichiometry, might pump $[Ca]_i$ down to its equilibrium value (20 μM) and that the SR or other intracellular Ca-accumulating mechanisms could then decrease the $[Ca]_i$ from this value to one below the contractile threshold. A number of difficulties exist with any scheme such as this. The fact that Na/Ca exchange is a reversible carrier movement means that the moment $[Ca]_i$ falls below the equilibrium value given above, the Na/Ca exchange would begin to pump Ca inward instead of outward; it would require an enormous energy expenditure by the SR to bring $[Ca]_i$ to values compatible with relaxation.

During the plateau of an action potential, Na/Ca exchange is not pumping out Ca. The SR Ca pump is reducing $[Ca]_i$. If it reduces it enough, the fiber will relax during the plateau. Repolarization initiates Ca extrusion by Na/Ca exchange and, depending on the properties of the two pumping systems, may even remove Ca from the SR by lowering $[Ca]_i$ to levels at which the SR pump is ineffective. The SR leak then supplies Ca to the Na/Ca pump.

Experiments with mammalian myocardium have been carried out to test the relative abilities of the SR and the Na/Ca exchange mechanism to control $[Ca]_i$. As shown in Fig. 6.9, the conclusion is clear. Caffeine-induced Ca release from the SR will only produce contraction if Na/Ca exchange is disabled by removing Na_o. Other

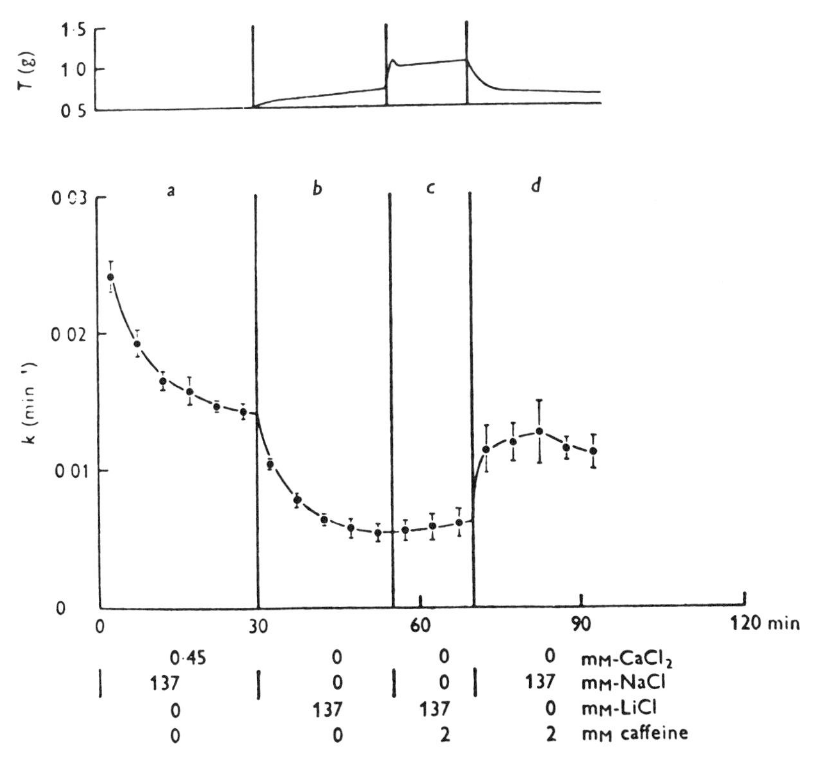

FIG. 6.9. Top: Plot of tension versus time. **Bottom:** Rate constant for ^{45}Ca loss from auricular trabeculae in Na Ringer's, Li Ringer's, and in both these solutions when 2 mM caffeine is applied. Note that tension is present only when Na/Ca exchange is disabled, while increased Ca efflux occurs only if external Na is present. (Data taken from ref. 1, Chapter 7.)

experiments show that contractility of all cardiac fibers is quickly lost if Ca_o-free solutions are applied. The conclusion drawn from such a finding is that the Na/Ca extrusion system can pump down $[Ca]_i$ to such low levels that stored Ca is lost. At constant heart rate, therefore, the stored Ca must be in a steady state so that the amount released during contraction equals the amount reaccumulated during relaxation.

Tension must be expected to vary in a cardiac fiber with $[Ca]_i$, as shown in Fig. 6.10. Note that tension is extremely small at 100 nM and about half-maximal at 1,000 nM. In squid axons, $[Ca]_i$ has been measured at 20nM; there is every reason to believe that the cardiac Na/Ca pump is capable of reaching such a low value. If such a value is reached, clearly the $[Ca]_i$ of the fiber is far from the contractile threshold. More important, the Ca concentration of the SR is five times lower than it would be if $[Ca]_i$ were 100 nM. As noted in Chapter 4, $([Na]_o)^4[Ca]_i = ([Na]_i)^4[Ca]_o(K)$. Since neither $[Na]_o$ nor $[Ca]_o$ change under physiological conditions, decreases in $[Na]_i$ must produce large changes in $[Ca]_i$. Such changes will greatly inhibit the rate at which Ca can enter with depolarization by Na/Ca exchange since this depends on $([Na]_i)^4$. The following are consequences of a low $[Ca]_i$ (one that might be produced

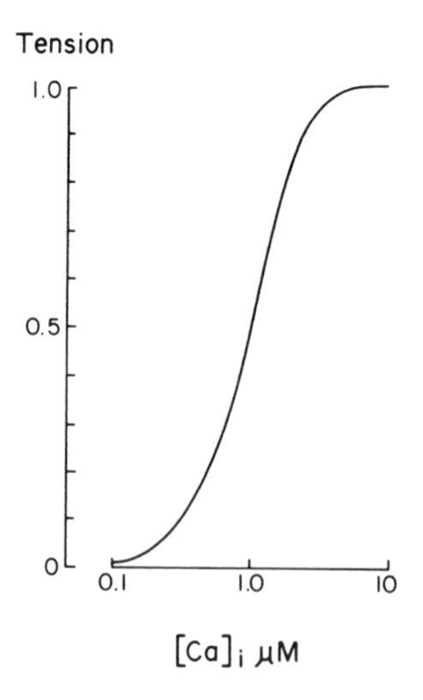

FIG. 6.10. Plot of tension developed by muscle as a function of ionized Ca in the fiber. Based on a variety of studies, as detailed in the review by Chapman (ref. 3, Chapter 5).

by preventing cardiac fibers from beating): (a) loss of SR Ca, (b) decline in $[Na]_i$, and (c) decline in inward Ca movement via Na/Ca with depolarization.

On the other hand, if $[Ca]_i$ were kept high but below the threshold for mechanical activation (i.e., 100nM), then the quantity of Ca in the SR would be five times that considered above; $[Na]_i$ would be higher; and depolarization would lead to an enhanced Ca entry via Na/Ca exchange and an enhanced release of Ca from the SR simply because it now contained more Ca.

How might these differing values for $[Ca]_i$ be produced? If a fiber is not contracting, the Na/K pump will pump down $[Na]_i$ to a value where pump efflux is equal to Na inward leak. As $[Na]_i$ declines, so does $[Ca]_i$; thus the lowest possible value for $[Ca]_i$ is reached. Measurements of Na entry in resting and contracting cardiac fibers (15) suggest that the fluxes are 15 and 55 pmoles/cm^2. Therefore, there is about a fourfold decrease in the rate of Na entry in a "resting" fiber over one that is contracting. Thus $[Na]_i$ would differ by a factor of 4 in the two kinds of fibers; and the consequences of this altered $[Na]_i$ are reflected in an altered $[Ca]_i$ and in the amount of Ca entering via Na/Ca exchange if a test depolarization is given.

To illustrate this point further, Fig. 6.11 shows the assumed $[Na]_i$ in a resting fiber (10 mM) together with the appropriate values for $[Ca]_i$ and Ca carrier current. When the fiber is stimulated to contract, $[Na]_i$ gradually rises to a new steady-state value (here assumed to be 15 mM) and $[Ca]_i$ to its appropriate equilibrium value (diastolic), a 1.5 increase in $[Na]_i = (1.5)^4$ or a fivefold increase in $[Ca]_i$. Ca rises also in the SR and in carrier current. There is a much more rapid loading of the SR from the Ca entry produced with each action potential; such a loading depends on the kinetic characteristics of the SR Ca pump in comparison with the Na/Ca exchange pump. As the frequency of stimulation is further increased, there will be insufficient time for the SR to reload with Ca between beats, and the content of stored Ca will fall. Unfortunately, the necessary experiments have not been done to characterize this pump interaction.

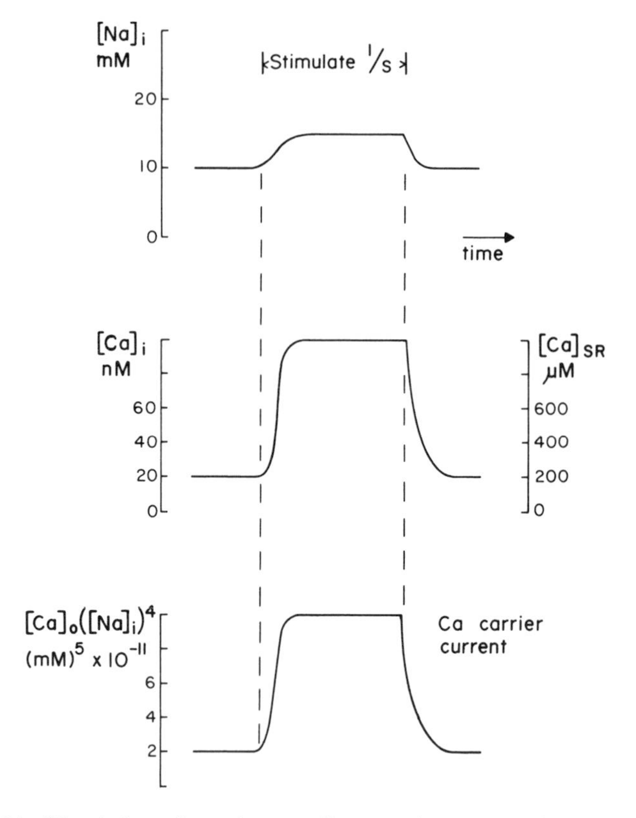

FIG. 6.11. Stimulation of a quiescent fiber can be expected to cause $[Na]_i$ to increase to a higher value. This value will be a function of frequency of stimulation. This change in $[Na]_i$ can be calculated to transform $[Ca]_i$ (*second tracing*). Since it is assumed that there is a constant relationship between $[Ca]_i$ and $[Ca]_{SR}$, the second trace also gives the change in concentration of Ca in the SR that must be expected from the change in $[Na]_i$. *Third trace,* expected carrier current of Ca (produced by Na/Ca exchange running backward). This too is a function of the product of $[Ca]_o$, and $[Na]_i$, as shown on the ordinate of this trace.

Despite its many unknown features, the SR Ca pump can be expected to behave as other carrier-mediated systems in that it must have a V_{max} (a limiting pump flux at saturation) and a K_m or a concentration of Ca where pumping is half-maximal. A comparison of the SR Ca pump with the sarcolemmal Ca pump (Fig. 6.9) shows that a functioning surface membrane Na/Ca exchange can prevent Ca release from the SR from inducing contraction, provided that E_m is maintained at resting values. This suggests that V_{max} is greater for Na/Ca than for the gated SR Ca release and that the SR Ca pump, which is ATP driven, must be slower still (or it would prevent gated Ca release from increasing $[Ca]_i$).

With respect to relative K_m values, if that for the SR is much less than that for Na/Ca, then the Na/Ca pump only pumps the $[Ca]_i$ down to modest levels (~ 20 μM), and the SR continues $[Ca]_i$ reduction to levels below the contractile threshold.

Several factors argue against any such arrangement. First, as $[Ca]_i$ declines below the equilibrium value for Na/Ca exchange (in the example above, for a 2 Na:1 Ca coupling), the mechanism will reverse. It is an empirical observation that the further away from equilibrium, the greater the flux via Na/Ca exchange. Hence the SR pump will have to pump down the residual $[Ca]_i$ and pump down the Ca entering at an increasing rate from outside. Second, the regulatory role for $[Na]_i$ would be the reverse of what is observed experimentally. That is, lowering $[Na]_i$ would mean that it would be easier for the SR to pump out the Ca entering from Na/Ca exchange since there would be less of it. By contrast, experimental interventions that raise $[Na]_i$ (ouabain) increase contractility. Hence the K_m for the SR must be higher than that for Na/Ca exchange.

Another way of analyzing the relative K_m of the SR and surface membrane Ca pumps is to consider a fiber with a high $[Ca]_i$ (i.e., immediately after a contraction) and to make a sudden change to $[Ca]_o = 0$. Both pumps continue to reduce $[Ca]_i$; but whichever pump has the highest value for K_m will be the one to stop first (as $[Ca]_i$ becomes much less than K_m). If the surface membrane pump stopped first, the SR would retain its accumulated Ca. But experimental evidence is clearly in favor of the pumping out of Ca from the SR by the surface membrane pump (as judged by loss of contractility). Hence the SR pump has a lower V_{max} and a higher K_m than that of the surface membrane pump.

When Ca is reintroduced into a Tyrode solution bathing a cardiac cell previously kept Ca_o-free, the mechanical response of the fiber to test depolarizations is a series of increases in tension over several beats. Such observations are sometimes used to justify an "elaborate" SR arrangement, whereby Ca taken up by the SR is moved to another compartment before being released to the contractile machinery. A simpler explanation could lie in the assumption that V_{max} is not sufficient to recharge the SR from the Ca entry in a single beat.

The considerations mentioned above suggest that the diagram shown in Fig. 6.12 is appropriate to the cardiac fiber. It shows a V_{max} for A (the surface membrane Na/Ca pump), B (the SR pump), and C (mitochondria), as well as K_m values. The ultimate value of $[Ca]_i$ at which there is no net flux via Na/Ca exchange for a Na ratio across the membrane of 11 and a E_m of -80 mV is 0.2 nM; however, small but finite leaks across the surface membrane are assumed to hold $[Ca]_i$ at about 20 nM, the "ultimate" concentration set by the leak rate. Since $[Ca]_i$ over a normal cardiac cycle changes

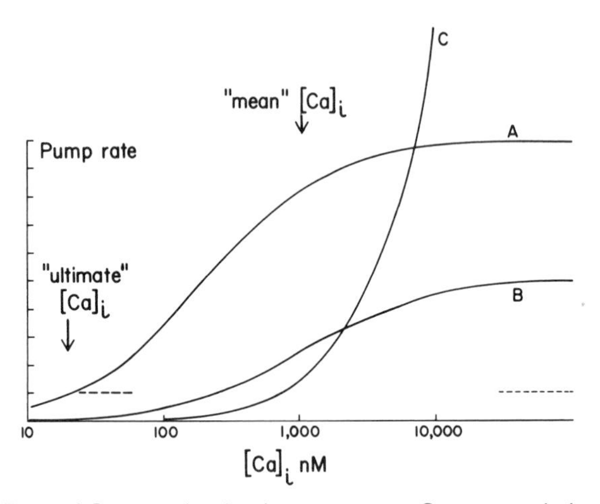

FIG. 6.12. Rate of Ca pumping by three separate Ca-accumulating systems as a function of $[Ca]_i$ in myoplasm. The [Ca] indicated by the arrow labeled "ultimate" $[Ca]_i$ is that set by voltage-sensitive leaks of Ca into the fiber. The concentration labeled "mean" $[Ca]_i$ is the time average [Ca] over a contraction. Part of Ca pumping will occur at a high rate and part at a low rate. A = Na/Ca; B = SR; C = mitochondria.

from values near the "ultimate" $[Ca]_i$ to several thousand nanomolar, a "mean" $[Ca]_i$ can be defined (Fig. 6.12) as the time average of the $[Ca]_i$. A value of 1,000 nM has been calculated for this, which is at the K_m of the SR; it is at the point where pump rate per change in $[Ca]_i$ is at its maximum.

If both the SR Ca pump and the Na/Ca pump operated throughout the cycle of cardiac contraction, both pumps would reach nearly maximal rates of operation as $[Ca]_i$ rose to contractile levels. By having Na/Ca electrogenic, the Na/Ca pump is stopped during the depolarizing phase of the action potential, and the SR pump has first claim on the available Ca. Repolarization then energizes Ca pumping at lower levels of $[Ca]_i$.

With two independently gated sources of Ca (the channel and Na/Ca pump) and a dependent gated source (the SR), there is a real possibility of oscillations in $[Ca]_i$ as a result of the variations in E_m. Such oscillations under various pathological conditions[1] are the result of an increased $[Na]_i$ (and hence $[Ca]_i$).

Mitochondria

Isolated mitochondria are also capable of concentrating Ca. Measurements indicate that the ratio of $[Ca]_i$ to $[Ca]_o$ may approach 10^3 when $[Ca]_o$ is about 1 μM. *In vivo* measurements with mitochondria in squid giant axons show little or no Ca accumulation when the $[Ca]_i$ of axoplasm is of the order of 20 to 100 nM. Little Ca is contained in mitochondria and releasable by FCCP or CN. Strong buffering of Ca_i by these organelles commences at a $[Ca]_i$ of about 1 μM. For such concentrations, only 1/1000th of the entering Ca into squid axons goes to increase $[Ca]_i$. Such findings strongly support only a secondary role for mitochondria in con-

[1]The current is described as TI (for transient inward) and results from cardiac glycoside intoxication. This can be expected to lead to increased $[Na]_i$ which in turn will lead to an increase in diastolic $[Ca]_i$. The enhancement in the concentration of both these ions means that I_C will be larger than normal for both inward and outward currents.

trolling [Ca] in the axoplasm or myoplasm of cells. On the other hand, [Ca] in muscle goes above the micromolar concentration range. Mitochondria tend to limit excursions in $[Ca]_i$ in contractile tissue. However, they do not participate in the kinds of Ca entry and release mechanisms that can be ascribed to the SR and to the Na/Ca exchange mechanism.

These conclusions about mitochondria have been reached from experiments on squid giant axons, such as that shown in Fig. 6.13. An axon was injected with both aequorin and phenol red (to confine $[Ca]_i$ measurement to the periphery); an increase in Ca influx then was made by increasing $[Ca]_o$ from 1 to 37 mM. The result was a large, transient entry; when $[Ca]_i$ rose high enough, mitochondrial

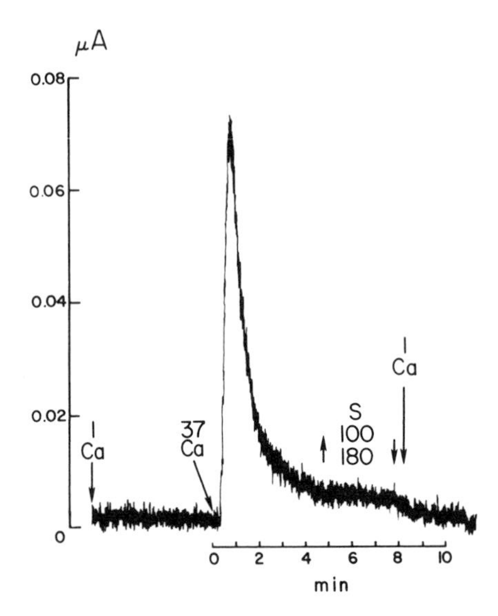

FIG. 6.13. Squid giant axon injected with both aequorin and phenol red (to confine light measurement to the axon periphery) is treated with 37 mM Ca sea water at the point shown by the arrow. The large transient can be shown to be due in part to mitochondria that are triggered to take up Ca only when $[Ca]_i$ reaches levels of the order of 1 μM. A stimulation (100 sec^{-1} for 180 sec) allows the entry of 10 times more Ca than that produced by a change from 1 mM to 37 mM Ca sea water. Mitochondrial Ca buffering, however, is such that there is virtually no [Ca] change. (Data taken from ref. 16.)

buffering was triggered. This buffering continued when $[Ca]_i$ was at a low level, since stimulation (with the entry of 10 times as much Ca) only slightly changed $[Ca]_i$.

A reasonable summary of how Ca is handled in internal compartments is that gating releasing Ca from the SR is probably slower than gating of surface membrane Ca channels, and reactivation of the SR gating mechanism is probably quite slow indeed. It is almost certain that the SR pump runs continuously, and that its rate of Ca reaccumulation depends on $[Ca]_i$. By contrast, Ca extrusion by Na/Ca exchange is stopped during the plateau; thus the SR pump has first claim on the available Ca. Upon repolarization, the Na/Ca pump may actually withdraw Ca from the SR as it pumps $[Ca]_i$ down to low levels. Ca pumping by mitochondria or its buffering by organic compounds must be expected to be only minor regulators of $[Ca]_i$ during the cardiac cycle.

The diagram in Fig. 6.12 has been drawn to show mitochondrial Ca pumping as a curve with a high value for K_m (Ca accumulation is very small below 1 μM Ca) but a curve with a very high V_{max}. This conclusion is derived from work with mitochondria studied both *in vitro* and *in vivo* in squid axons. In studies of the most peripheral mitochondria in squid axons using aequorin as a Ca indicator, however Mullins and Requena (16) found that mitochondria did require about 1 μM Ca to start Ca accumulation; once started, this accumulation could continue until Ca reached very low levels. Furthermore, a stimulation of the fiber that is known to introduce 10 times more Ca than the change in $[Ca]_o$ results in virtually no change in $[Ca]_i$. This implies that the mitochondria, once primed, can take up Ca at the very low levels of concentration shown on the record.

Finally, compounds such as ATP can complex Ca or Mg with equal facility; since MgATP is a substrate for Na/K pumping and ATP is present in millimolar concentrations in myoplasm, most of the Mg will be complexed with such compounds. With a large Ca load, since the affinity constants for Mg and Ca with ATP are virtually identical, CaATP and MgATP would be formed in concentrations reflecting the relative concentrations of these two ions.

It is unlikely the CaATP is ever formed in significant amounts under physiological conditions; clearly, however, ATP and other compounds are capable of buffering the concentration of this ion.

REFERENCES

1. Deitmer, J. W., and Ellis, D. (1978): *J. Physiol. (Lond.)*, 277:437–453.
2. Post, R. L., and Jolly, P. C. (1957): *Biochim, Biophys. Acta*, 25:118–128.
3. DiPolo, R. (1978): *Nature*, 274:390–392.
4. Biedert, S., Barry, W. H., and Smith, T. W. (1979): *J. Gen. Physiol.*, 74:479–494.
5. Sjodin, R. A.(1980): *J. Gen. Physiol.*, 76:99–108.
6. Mullins, L. J., and Noda, K. (1963): *J. Gen. Physiol.*, 47:117.
7. Mullins, L. J. (1977): *J. Gen. Physiol.*, 70:681–695.
8. Baker, P. F., Blaustein, M. P., Hodgkin, A. L. and Steinhardt, R. A. (1969): *J. Physiol. (Lond.)*, 200:431–458.
9. Hiraoka, M., and Hecht, H. H. (1973): *Pfluegers Arch.*, 339:25–36.
10. Isenberg, G., and Trautwein, W. (1974): *Pfluegers Arch.*, 350:41–54.
11. Isenberg, G., and Trautwein, W. (1975): *Pfluegers Arch.*, 358:225-234.
12. Gadsby, D. C., and Cranefield, P. F. (1979): *Proc. Natl. Acad. Sci. USA*, 76:1783–1787.
13. Ellis, D. (1977); *J. Physiol. (Lond.)*, 273:211–240.
14. Fabiato, A., and Fabiato, F. (1979): *Annu. Rev. Physiol.*, 41:473–484.
15. Langer, G. A (1964): *Circ. Res.*, 15:393–405.
16. Mullins, L. J., and Requena, J. (1979): *J. Gen. Physiol.*, 74:393–413.

Chapter 7

Changes in Transport During the Action Potential

CYCLIC NATURE OF [Ca] CHANGE

In muscle, $[Ca]_i$ must change from a "resting" value that is below the threshold for mechanical activation (20 to 100 nM) to a value in the range of 10^3 to 10^4 nM, depending on the extent of contraction to be produced. Although higher values of $[Ca]_i$ are possible, they produce no further mechanical response. The cardiac fiber has a peculiar action potential: it has a long duration that can be up to half the duty cycle. It has been claimed that this long duration action potential is to ensure refractoriness; a more important role is that of reversing Na/Ca exchange.[1] This change both stops Ca efflux and initiates Ca influx, thus providing the fiber interior with a source of Ca that can complement the Ca derived from Ca^{2+} moving through slow inward channels or the Ca released from internal stores. Each of these sources of Ca is subject to its own regulatory control (for example, norepinephrine for I_{si} and $[Na]_i$ for Na/Ca), so that a modulation of both beat duration and strength is continuously possible. The change of Ca flux into the myoplasm of a cardiac fiber is brought about because the change in membrane potential during the action potential (a) opens excitable, gated, channels to which Ca is permeable, and (b) changes

[1]I have been reminded on more than one occasion by Paul Cranefield that the last thing a working myocardium needs is a twitch. The arrangements for contraction in heart require a "slowly beginning but long sustained tension needed to pump blood into the highly elastic arterial tree."

the energy of the Na electrochemical gradient ($E_m - E_{Na}$) from about -140 mV to about -60 mV; this latter value is compatible with an elevated $[Ca]_i$ in the range where muscle contraction takes place.

REQUIREMENTS OF THE STEADY STATE

In a heart beating at a constant rate, the cycle of $[Ca]_i$ change outlined above can take place only if all ions that move across the membrane with excitation are in a steady state from beat to beat. This means that the entry and exit of any ion must be equal, or the concentration of the ion integrated over a beat interval must remain constant. We have seen how the Na/K and Na/Ca transport systems must be integrated; here it is proposed to examine how Ca can be in a steady state over the duty cycle of the heart. Recognized sources of Ca are (a) the SR, (b) Ca current via channels, and (c) Ca entry via Na/Ca exchange.

The SR must undergo a closed cycle, whereby the amount of Ca it releases per beat is taken up by the SR; were this not so, the fiber would not be in a steady state. Ca current entering by channels must be reversed by the expenditure of energy via pumping, since peak values for $[Ca]_i$ during maximal contraction are of the order of 10 μM while $[Ca]_o$ is 2 mM. This is an $E_{Ca} = +76$ mV; the plateau would have to be more positive if Ca were to flow outward. The only source of outward Ca movement is the Na/Ca exchange or other substrate-driven Ca pumps. This outward movement must reverse the entry of Ca not only by Ca channels, but also via Na/Ca exchange. Thus

$$\int_{t=0}^{t=p} [I_{Ca} + (-I_C)]\, dt = \int_{t=p}^{t=d} I_C dt$$

where $t = p$ is the duration of the plateau, $t = d$ is the duration of the diastolic interval, I_{Ca} is the channel current, and I_C is the Na/Ca carrier current. Note that I_C and the Ca current carried by the carrier have opposite signs but equal magnitudes. The relationship in the simplest case is shown in Fig. 7.1. The carrier current is

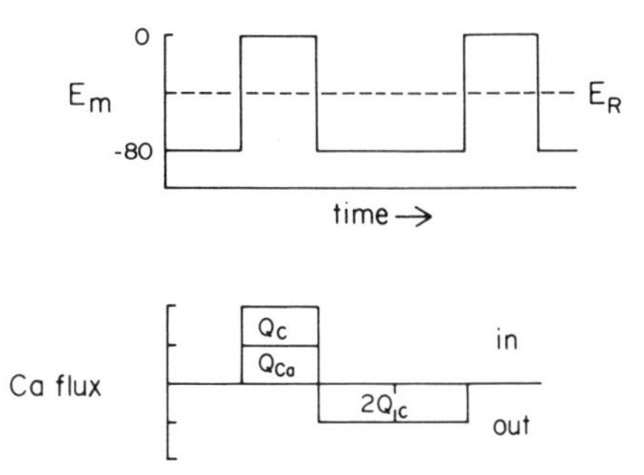

FIG. 7.1. Membrane potential of a cardiac fiber is assumed to vary as shown (*upper trace*). *Dashed line,* E_R, is the assumed reversal potential of the carrier current. Because the membrane is depolarized 40 mV positive to E_R for one-third of the time and hyperpolarized 40 mV negative to E_R for two-thirds of the time, it is assumed that Q_C, the charge transferred inward by a carrier-mediated Ca movement during depolarization, is balanced by an equal exit of Ca during repolarization. Because repolarization generates $2Q_C$, the charge transfer via the Ca channel, Q_{Ca}, is assumed to equal Q_C and to be reversed during repolarization.

constant in amplitude as a function of time. This simplified analysis suggests that Ca delivered by Na/Ca and by Ca channels are of the same order of magnitude. Conditions in the cardiac cell can be manipulated so that Na/Ca does not itself deliver much Ca; still it must be present to reverse the Ca gain from Ca channel currents.

VELOCITY OF Na/Ca EXCHANGE

The analysis just presented indicates that Ca extrusion cannot be a small fraction of I_{si}. This current flows for at least 30% of the duty cycle of a cardiac action potential and must be reversed during the diastolic interval. Some cardiac action potentials last for 50% of the duty cycle and therefore set a minimum value for Na/Ca exchange of a Ca flux at least equal to Ca channel current. Additional Ca leaks and the Ca entry from the Na/Ca system running backward would make a realistic maximal value for Ca extrusion at least equal to twice I_{Ca}.

Other important experiments show that if guinea pig auricles are treated with caffeine in Na Ringer's (1), there is a rise in Ca efflux but no contraction. If the same experiment is repeated in Na-free Ringer's, there is a substantial contraction but no rise in Ca efflux. These observations are interpreted by assuming that the velocity of outward transport of Ca from mammalian cardiac fibers is much greater than Ca entry via either slow channels or Ca release from the SR.

Further experiments with this preparation (1) show that the rate constant for Ca efflux can rise more than 10-fold by treatment of the auricles with CN in Na Ringer's (presumably a release of Ca from mitochondria) with no detectable increase in fiber tension. The removal of Na in Ringer's leads to a fall of Ca efflux to control levels and an abrupt rise in tension. Again, these measurements indicate that Na/Ca exchange (even in the absence of ATP) is capable of keeping $[Ca]_i$ at levels below the threshold for contraction. An inhibition of this pumping leads to the immediate production of tension.

It must be emphasized that all measurements in cardiac tissues are consistent with the idea that Na/Ca exchange is a system that can hold down $[Ca]_i$ to low levels (at least to levels below those required for contraction), regardless of the source of Ca for contraction. It is only when the membrane potential or the Na gradient is abolished or reduced that contraction can occur. The analysis so far presented indicates that both these operations specifically inhibit or reverse Na/Ca exchange.

CHANGES IN $[Ca]_i$

The rapid upstroke of a normal cardiac action potential results from the rapid regenerative entry of Na via Na channels. This leads to a change in potential of about -85 to $+5$ mV, which in turn allows an entry of Ca in exchange for Na_i via Na/Ca exchange. More slowly, the *si* channels of the membrane open, and a current carried in part by Ca begins to flow. A signal is passed to the SR that causes a release of Ca. This increase in Ca in the myoplasm leads not only to contraction but to an altered flow of carrier and channel currents. In a modest contraction where $[Ca]_i$ rises to perhaps 1 μM, the value of E_{Ca} reaches +100 mV. If E_{Na} is +50 mV, then $E_R = (2 \times (50) - 100)$ or 0. This is close to the plateau, so that Ca entry via Na/Ca exchange will cease. For the channel current, $I_{Ca} = g_{Ca}(E - E_{Ca})$. This will change from (0 − 140) to (0 − 100), or a relatively small change. It may be concluded that changes in $[Ca]_i$ involved in contraction will substantially affect Na/Ca and the resultant requirement for flux balance.

CARRIER CURRENTS

Figure 7.2 is a current-voltage plot for Na/Ca carrier current at two separate values for E_{Ca}: the first for the normal value of $[Ca]_i$ in a resting fiber (E_{Ca} = +140 mV) and the second for the case where Ca entry via *si* channels, via SR Ca release, and via Na/Ca carrier-mediated Ca entry has changed E_{Ca} to +100 mV. The diagram has also been drawn for a particular value of E_{Na} of +40

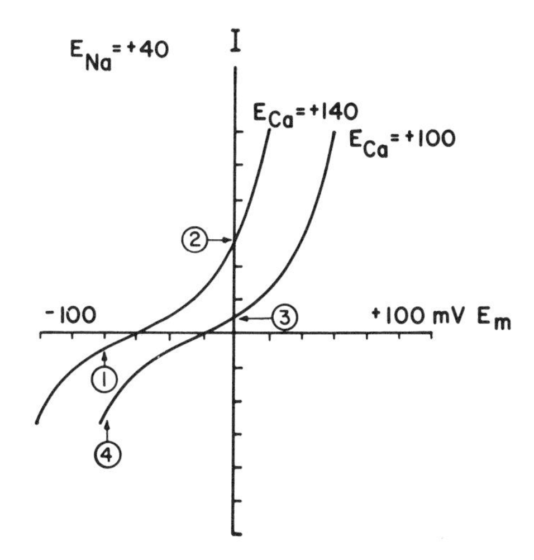

FIG. 7.2. Current-voltage diagram for cardiac fiber generating a carrier current by Na/Ca exchange. E_{Na} is fixed at +40 mV. The resting membrane potential is assumed to be −80 mV, the plateau to be zero membrane potential, and the change in $[Ca]_i$ to involve the change of E_{Ca} from +140 to +100. (See text.)

mV. Hence at a resting membrane potential of -80 mV and E_{Ca} of $+140$, the value of E_{R} is -60. At the resting potential there is a net extrusion of Ca (point *1*). When an action potential is initiated, E_m is moved to about zero (point *2*) (a change of carrier current from a small inward one to a large outward one). The ensuing Ca release from internal stores plus Ca entry from outside leads to a change in E_{Ca} to $+100$. Thus Ca movement by Na/Ca transport is now governed by point *3*. Finally, the resting potential is recovered, so that point *4* is the proper value for Na/Ca transport. The SR and other processes (Na/Ca pumping) reduce $[\mathrm{Ca}]_i$, so that E_{Ca} moves gradually to $+140$ mV (its initial value) and hence returns to point *1*. This cycle indicates that there are both transient and steady state values for Na/Ca transport calculated on the assumptions that (a) E_{Na} remains constant over the action potential cycle, and (b) E_{Ca} and E_m vary because of both Ca entry and internal release and because E_m is varied mainly by transient changes in P_{Na}, P_{K}.

This cyclic change in Ca entry and exit via Na/Ca exchange imposes some important limitations on speculations regarding the process. Any assumptions made must conform to the requirement that Ca fluxes balance over the cycle.

To illustrate the sensitivity of this requirement, Fig. 7.3 has been drawn to examine the responses shown in Fig. 7.2 as a function of time for three values of E_{Na}. The same values of E_{Ca} and its change during the plateau are assumed. The results can be summarized as follows: if $E_{\mathrm{Na}} = +30$ mV, then the time integral of Ca extrusion is inadequate to reverse even the entry of Ca via Na/Ca exchange, much less other entries of Ca via I_{si} or other leaks. Transiently, this value of E_{Na} allows large Ca entries, possibly to recover intracellular Ca that might have been lost; but no steady-state condition is possible. For $E_{\mathrm{Na}} = +40$, there is a symmetrical response: substantial Ca efflux during the diastolic interval and a gain during systole. This is a possible value for E_{Na}. When $E_{\mathrm{Na}} = +50$, there is a requirement for massive Ca entries via channel and leak to supply enough Ca for pumping purposes. This Ca entry must come from non-carrier-mediated Ca movement. Thus Ca pumping is highly sensitive to $E_{\mathrm{Na}} - E_{\mathrm{Ca}}$, with a probable value of about 100

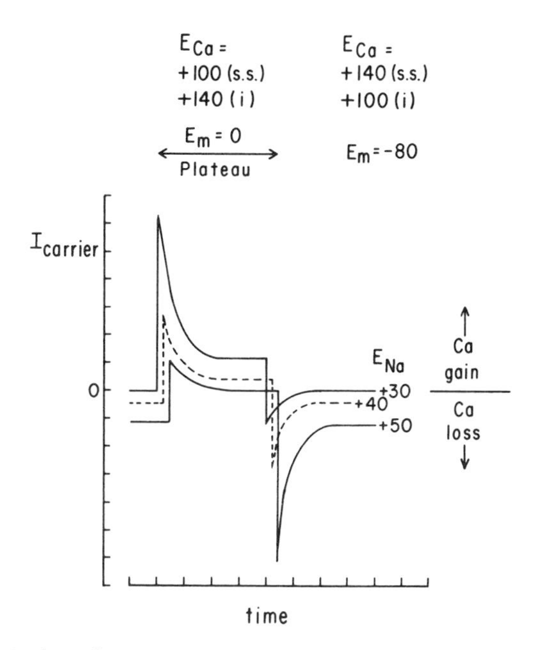

FIG. 7.3. Plot of carrier current via Na/Ca exchange versus time for three values of E_{Na}, as indicated on the individual curves. Note that each curve has been displaced slightly to the right in order to avoid complete overlapping. The curves start prior to the plateau. The time of the plateau is indicated by the period encompassed between the arrows. The change in E_{Ca} during the plateau and during the diastolic interval is indicated by the legends at the top of the figure. The instantaneous (i) and steady state (s.s.) values are shown for both cases. The currents have been computed on the basis of a sinh relationship between current and displacement of the membrane potential from E_R.

mV during diastole. A change of ± 10 mV has a decisive effect on overall Ca flux balance.

CONTRIBUTIONS OF Na/Ca EXCHANGE TO THE ACTION POTENTIAL

The assumption has been made that Na/Ca exchange is influenced by changes in membrane potential produced by various conductance changes but that the exchange itself does not produce sufficient membrane current (I_C) to affect E_m. This section addresses the validity of this assumption.

The rising phase of the cardiac action potential is produced by I_{Na} flowing at a rate sufficient to produce a change of membrane potential of as much as 800 V/sec. As the membrane potential passes -40 mV, the assumed reversal potential of I_C, it activates an outward current of Na (and an inward movement of Ca); when E_m reaches zero, the magnitude of this current is at most a few microamps per square centimeter, but I_{Na} (initially 800 $\mu A/cm^2$) is rapidly declining as E_m approaches E_{Na}, both because inactivation is taking place and because of a decrease in driving force. When E_m approaches $+25$ mV, the driving force for I_{Na} is greatly reduced, and inactivation has acted to reduce the current. Thus it is not unreasonable to suppose that I_C may play a significant role in the generation of the transient outward current, I_{qr}, that occurs at the onset of the action potential.[2]

This current has had a long history. It was originally suggested that it was a Cl current, since removal of this ion from the external solution greatly reduced the current. Kenyon and Gibbons (2), however, found that replacements for Cl often altered the free Ca in the bathing solution. When this change was considered, Cl removal had less of an effect on the measured current, while 4-aminopyridine, an agent that poisons K channels, reduced I_{qr} substantially.

Finally, Siegelbaum and Tsien (3) have shown that Cl removal has a large effect in reducing the current, even when Ca complexing is considered. The authors also show that substances known to block I_{si}, such as D600 or Mn^{2+}, also abolish I_{qr}. The injection of EGTA also abolishes I_{qr}. This rather confusing array of experimental findings can be simplified somewhat by assuming that the Kenyon and Gibbons (2) results included currents from voltage-sensitive K channels as well as the Cl_o-sensitive mechanism.

[2]This occurs only in Purkinje fibers, presumably because the repolarization process is much delayed as a result of large fiber size and consequently the requirements for a long time to raise $[Ca]_i$ levels where it can promote the opening of K channels. In auricular and ventricular fibers there is possibly a greater temporal overlap of I_C and I_K.

The removal of Cl_o has been expected to abolish the Cl gradient across the membrane. If this is responsible for H^+ transport, an internal pH change in the cell can be expected. The subject is complex, since there are transient and steady changes and a dependence on the nature of the anion, but an acidification produced by CO_2 has been shown by Lea and Ashley (4) in barnacle fibers to lead to an increase in $[Ca]_i$, a change that will abolish or decrease I_C. The injection of EGTA into a fiber can also be expected to produce a substantial acidification, as it should be capable of withdrawing from the SR and complexing substantial quantities of Ca. In the absence of adequate pH buffering, this substance must be expected to decrease pH_i.

The use of such substances as Mn^{2+}, while not specific for any particular ion transfer system, has been shown to block both I_{si} and Na/Ca exchange. The expected effect of such a substance is to block I_C and hence I_{qr}; indeed, such an effect is observed.

An important demonstration by Siegelbaum and Tsien (Fig. 7 in ref. 3) is the close correspondence between the magnitude of I_{qr} and the development of tension. Both current and tension start at -40 (the reversal potential of I_C) and rise steeply with depolarization. Another similarity is that I_{qr} inactivates with depolarization and that I_C must be expected to inactivate as $[Ca]_i$ rises. Both these effects are expected from the assumption that it is the Na/Ca carrier that is introducing Ca into the fiber before significant current flow via I_{si} has commenced. The major fraction of the plateau is undoubtedly dominated by the flow of I_{si}, while the continuously rising $[Ca]_i$ ultimately leads to the activation of Ca_i-sensitive K channels that finally open and allow an abrupt repolarization of the fiber to resting levels of membrane potential.

Another place in the action potential where Na/Ca may contribute electric current that significantly affects the membrane potential is the generation of pacemaker current in Purkinje fibers. McAllister and Noble (5) showed that this current could not be recorded in the absence of Na_o, while Isenberg (6) showed that the increased K conductance, leading to the repolarization from the plateau to the resting potential, depended on $[Ca]_i$. A conclusion from these two findings is that pacemaker current is in fact controlled by I_C, the Na/Ca exchange current, that flows inward as Ca is pumped from the fiber. This inflow of Na removes Ca_i, thereby lowering $[Ca]_i$. This in turn lowers membrane K conductance. Note that $[Ca]_i$ can be expected to decline exponentially with time. Because pacemaker current is generated by a carrier-mediated process, it can be expected to have a large temperature coefficient (i.e., unlike channel-mediated conductances).

Previous reports about the nature of this current suggested that it was a gated K current. This conclusion resulted mainly from a demonstration that the reversal potential was near E_K and far from E_{Na}. The introduction of a current-generating system with a reversal potential near -40 mV and recognition that the pacemaker process should be a mixture of inward carrier and outward K current make reversal potential measurements understandable on an other-than-K basis.

A second point is that the reversal potential for pacemaker currents and the range of potentials where they are activated is clearly different in Purkinje fibers and in sinus and atrial fibers. This difference is understood by the supposition that the current itself is a result of a carrier Na/Ca current whose reversal potential is dependent on Na_i, Ca_i, rather than on a channel that must have different characteristics in different tissues. Many treatments have as a common basis an increase in $[Ca]_i$ (epinephrine, depolarization). The effects of these treatments on pacemaker current are compatible with the idea that $[Ca]_i$ gates a K current and that it is the inward current associated with the pumping out of this increased $[Ca]_i$ that is the requirement for the pacemaker process.

Figure 7.4 is a diagram of an action potential in cardiac Purkinje fibers. Two regions of the potential-time diagram are likely to be governed almost exclusively by Na/Ca exchange currents, while other parts are governed by previously described currents. These regions are the current I_{qr} that is involved in the return of the potential from near E_{Na} to the plateau and the pacemaker potential responsible for the depolarization of the membrane potential to the threshold for the activation of I_{Na}. The figure has been drawn to

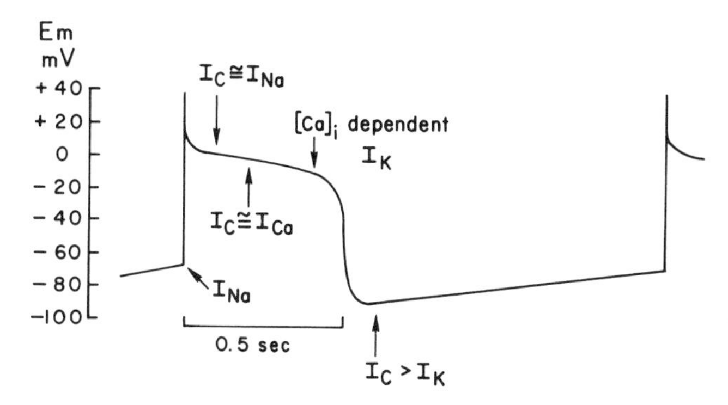

FIG. 7.4. Purkinje fiber action potential. I_{Na} is considered to bring E_m close to E_{Na}. Inactivation of I_{Na} is such that early in the plateau there is a point where $I_C = I_{Na}$. Further stabilization of the plateau is achieved because I_{Ca} has increased as I_C decreases; thus a point is reached where $I_C = I_{Ca}$. The increase in $[Ca]_i$ opens K channels. Membrane repolarization makes I_C large and inward, and the decrease in [Ca] from Ca pumping reduces $g_{K(Ca)}$, thereby allowing pacemaker depolarization.

show an action potential initiated by a large increase in I_{Na}, with the potential change producing a small, instantaneous current, I_C, that equals I_{Na} only when this has substantially inactivated ($I_C = I_{Na}$). A bit later in the plateau, I_{Ca} has risen and I_C declined so that these are both equal; at the end of the plateau, $[Ca]_i$ has risen sufficiently so that a Ca_i-activated increase in g_K is initiated, leading to a large I_K that repolarizes the membrane. This change in membrane potential now reverses the direction of I_C so that $[Ca]_i$ declines. This decline reduces g_K so that the inward I_C is capable of producing a slow, steady decrease of E_m that continues until the threshold for the initiation of a Na current is reached.

Purkinje fibers are not normally the pacemaking cells of the heart. This function is carried out by the SA node, whose cells have a low (~ -55 mV) membrane potential. This existence of such a low membrane potential means a high $[Ca]_i$ for the cell and hence that Na/Ca exchange and an inward I_C will be highly activated during diastole. The nodal action potential is brought about not by Na entry via Na channels but by Ca entry via Ca channels; the low membrane potential means that little Ca will be introduced by the mechanism shown in Fig. 7.3 but that there will be large inward carrier currents on repolarization. These inward currents will oppose outward K currents so a reversal potential between E_K and E_R is to be expected.

In contrast, ventricular fibers do not show pacemaking activity. An explanation is needed of why this cell type is different. The most obvious explanation is that upon the return of the membrane potential to the baseline following an action potential, $[Ca]_i$ is very low, so that there is little Na/Ca current available. The behavior of the resting membrane potential also depends critically on the extent to which membrane K conductance is a Ca_i-dependent rather than a Ca_i-independent one.

CHANGES IN Na ENTRY

The concentration of Na in cardiac fibers is almost 10^6 that of $[Ca]_i$. Thus the entry of Na in connection with a single action potential cannot affect $[Na]_i$. In turn, this means that the rate of Na

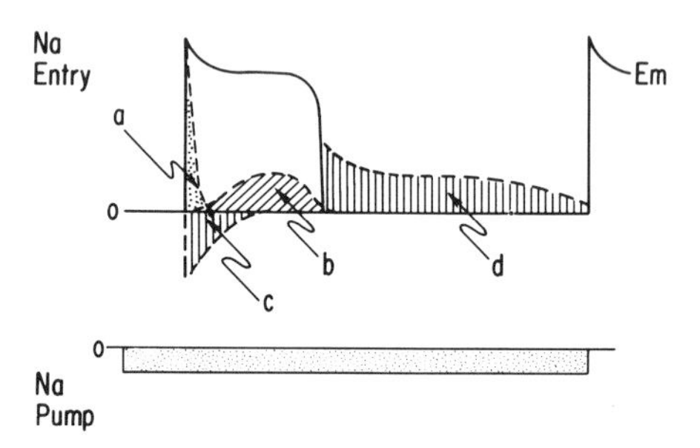

FIG. 7.5. Line describing the action potential of a cardiac fiber and the complex pattern of Na entry expected during the action potential and in the diastolic interval. The area encompassed by *a* is the capacitative Na entry necessary to bring about the depolarization; *b* is Na flowing through inward channels; *c* is an outflow of Na that occurs in a coupled manner due to Na/Ca exchange; and *d* is the Na entry necessary to reverse the Ca entry (not shown). *Below,* the Na pump is assumed to operate at a constant velocity and to produce a Na flux whose time integral is precisely equal to the sum of a + b + c + d.

pumping during an action potential will be constant, since plasma [K] maximally stimulates the pump. It is sometimes assumed (7) that Na entry is mainly via Na channels during the depolarizing phase of the action potential. The actual pattern of Na influx that occurs, however, is complex (Fig. 7.5). Figure 7.5 is drawn on the basis that Na entry via Na channels (a) is the minimum required to produce the depolarization observed and is only a fraction of total Na entry, which is 50 pmoles/cm^2 imp. The other components of Na entry are: (b) Na entry via slow inward channels (about 50% of Ca entry); (c) Na exit via Na/Ca exchange that occurs early in the action potential and before $[Ca]_i$ in the fiber increases (equal to Ca entry via Na/Ca); and (d) Na entry that occurs to pump out the Ca entry occurring during (b) and (c).

REFERENCES

1. Jundt, H., Porzig, H., Reuter, H., and Stucki, J. W. (1975): *J. Physiol.*, 246:229–253.
2. Kenyon, J., and Gibbons, W. (1977): *J. Gen. Physiol.*, 70:635–660.
3. Siegelbaum, S., and Tsien, R. (1980): *J. Physiol.*, 299:485–506.
4. Lea, T. J., and Ashley, C. C. (1978): *Nature*, 275:236–238.
5. McAllister, R., and Noble, D. (1966): *J. Physiol.*, 186:632–662.
6. Isenberg, G. (1977): *Pfluegers Arch.*, 371:71—76.
7. Morad, M., and Goldman, Y. (1973): *Prog. Biophys. Mol. Biol.*, 27:257–313.

Chapter 8

Depolarization and the Generation of Tension

SENSITIVITY OF CONTRACTION TO E_m

Lüttgau and Oetliker (1) have shown, in frog skeletal muscle, that at the foot of the tension-membrane depolarization curve tension increases *e*-fold per 3.5 mV of depolarization. By contrast, Chapman (2), in his review of excitation-contraction coupling, has tabulated *e*-fold increases in tension of cardiac muscle when the membrane is depolarized from 6.5 to 12 mV. A high sensitivity of a skeletal muscle to depolarization suggests that most of the Ca comes from the SR; a lower sensitivity of cardiac muscle means that Ca comes from other sources. In the case of the Na/Ca exchange mechanism, at equilibrium, a change in membrane potential of 12.5 mV is equal to an *e*-fold change in $[Ca]_i$. The response of the slow inward channel current to depolarization is similar to the above; if, as some mammalian cardiac fibers show, there is a sensitivity of 6.5 mV per *e*-fold increase in tension, less Ca comes from SR and most from other sources. This measurement does not differentiate between Ca influx via Ca channels or that via Na/Ca exchange; but other experimental findings, such as the relative absence of a sensitivity to Na-free conditions, suggest that a substantial amount of Ca may enter via slow inward channels. This does not mean that Na/Ca exchange is unimportant in mammalian cardiac fibers, because the change in membrane potential from resting levels to the plateau can be expected to stop Ca efflux. This is important in regulating the rate that $[Ca]_i$ can increase from the combined effects of SR Ca release and Ca entry via channels.

INITIATION OF CONTRACTION

Contraction of a cardiac muscle fiber depends on the introduction into the myoplasm of sufficient Ca to react with the contractile proteins present in the fiber. Three recognized sources of Ca may be involved in this process: (a) the Ca existing in an internal membranous store, the SR, (b) Ca entry via gated Ca channels, and (c) Ca entry via the Na/Ca exchange mechanism running in a reverse direction. The general arrangement of these various sources for Ca in cardiac fibers is shown in Fig. 8.1. In the SR, an ATP-driven pump removes Ca from the sarcoplasm, and a gated channel allows the release of Ca when it receives an appropriate signal. It is not clear from the experimental information available what produces the signal for the release of Ca from the SR. Many mechanisms have been proposed, and some are discussed briefly below. The main comment that can be made about the SR and its role in contractile processes is that in the case of fast skeletal muscle, it is clear that virtually all the Ca that is required for a twitch is released

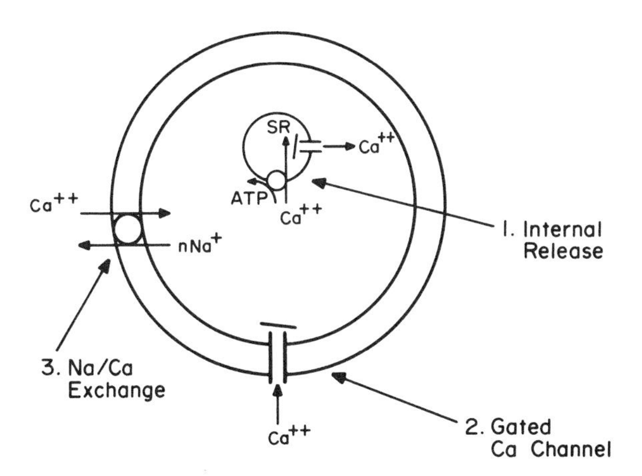

FIG. 8.1. The three recognized sources of Ca for the production of muscle contraction. *1:* Internal release via the SR consisting of a gated channel releasing Ca and a continuously running ATP-driven Ca pump. *2:* A gated Ca channel in the membrane that is voltage sensitive and responsible for I_{si}. *3:* Na/Ca running backward when E_m or E_{Na} is reduced.

from the SR when an action potential passes over the surface membrane and down the T-system; hence, a skeletal muscle fiber is substantially independent of the $[Ca]_o$. The opposite is the case for a cardiac fiber; in every case, contraction is lost upon reduction of $[Ca]_o$.

The second mechanism mentioned above (namely, that of a gated Ca channel) is relatively unimportant in skeletal muscle contraction but has been shown to introduce a large amount of Ca into cardiac fibers. The electrophysiology of the operation of this channel is not discussed here, since the treatment in this volume is limited to transport rather than electrophysiological processes. However, it is clear from studies that have been carried out that the channel is slow in opening compared with Na channels and correspondingly slow in inactivating; thus it is capable of introducing Ca into the fiber over times comparable to the duration of an action potential.

The third mechanism is that of Na/Ca exchange. It has long been known that removal of external Na in a Ringer's solution causes a transient contraction of frog cardiac muscle. This is the generally expected behavior of the Na/Ca exchange system shown to operate in cardiac muscle when E_{Na} is made negative (i.e., when the $[Na]_o$ is made lower than $[Na]_i$). This ionic change is impossible under physiological conditions. An analysis of the Na/Ca system shows that not only can it be reversed in its direction by a change in E_{Na}, but also the direction of net movement of Ca can be reversed by a change in the membrane potential. The section on the energetics of Ca transport indicates that changes in E_{Na} and E_m should produce equivalent changes in the direction of movement of Ca. Therefore, a change in E_m of cardiac fiber from resting membrane potential (approximately -80 mV) to plateau potential (possibly $+10$ mV) should substantially increase the entry of Ca.

SOURCES OF CONTRACTILE Ca

To what extent does the Ca that is actually used for the initiation of contraction come from one or the other of the sources indicated above? Answers to this question have generated substantial controversy for the last 10 years. The experimental operations necessary to identify sources of Ca for contraction are difficult to carry out. Cardiac fibers vary greatly in the extent to which they use the various sources of Ca classified above. How can one experimentally identify a particular source of Ca for a particular response of the contractile system? One way is to note that chemical reagents, such as caffeine, will release Ca from the SR without substantial effects on the Ca fluxes across the surface membrane.

Experiments done by Jundt et al. (3), using guinea pig auricles, have shown that the effect of caffeine in producing tension was obvious only when the external solution, which contained no Ca, also was Na-free. The conclusion is that if the Ca pumping system in the surface membrane is working (which occurs in the presence of Na_o), then as fast as Ca is released from the SR by the caffeine, it is pumped out of the fiber and has no opportunity to interact with the contractile machinery. The experiment suggests that the SR release of Ca is by no means the major source of Ca for the contractile machinery, and that pumping systems exist that are capable of rapidly disposing of the released Ca from the SR.

Another experiment measures the mechanical response of a cardiac fiber in a solution that is suddenly changed to Ca-free conditions. Under these circumstances, the tension that can be elicited from the fiber falls with the time constant by which Ca falls in the extracellular space. Again, this suggests that release from the SR which should not be affected by this change in external solution is not large enough to allow $[Ca]_i$ to rise to levels necessary for the initiation of contraction; instead, the pumping activities of the Na/Ca system are sufficiently strong to remove the Ca as it is released. A similar experiment carried out with the skeletal muscle fiber would show that tension was unaffected by this treatment with Ca-free solution. This observation emphasizes two points: (a) skeletal muscle has a more extensive SR and one more heavily loaded with Ca; and (b) skeletal muscle has substantially lower density of Na/Ca pumping sites; otherwise, these would pump out the internal store of Ca.

Figure 8.2 shows action potentials for skeletal and cardiac mus-

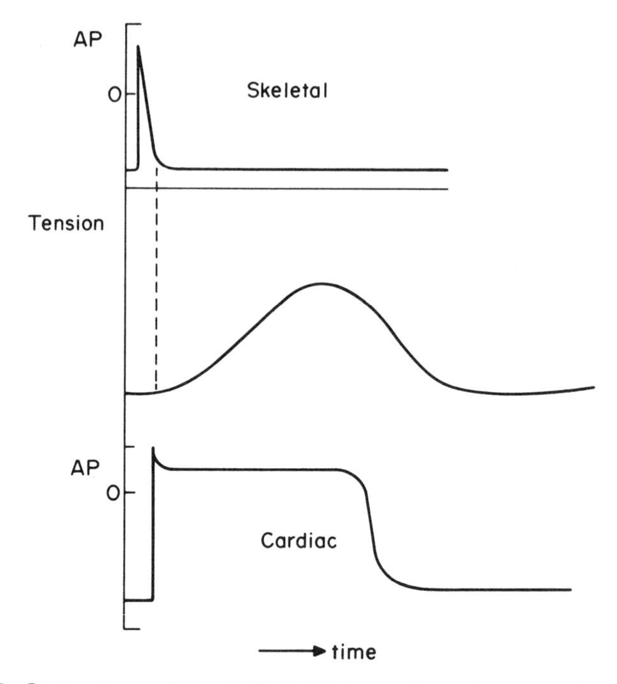

FIG. 8.2. Comparison of the fast skeletal muscle action potential and the subsequent development of tension with the cardiac action potential and its development of tension. In the first case, bioelectric activity is over before tension begins; in the latter, tension and membrane depolarization are much closer in phase.

cle, together with a diagram showing how the tension is related to the timing of the action potential. In skeletal muscle, tension begins to rise at a time when the action potential has been completed. In cardiac fibers, there is substantial variability in the shape of the action potential; as a general rule, it can be said that tension continues as the membrane depolarization stays at the level of the plateau, and that a repolarization of the fiber is a signal to produce relaxation. Cardiac and skeletal SRs may have substantially different operating mechanisms, but this seems unlikely from what we know of other types of membrane pumps and channels. The behavior of such things as Na/K pumps and Na channels is universal;

and it would be surprising indeed if the SR pump/channel mechanisms were different.

If the SR release mechanisms are similar in skeletal and cardiac muscle, all the SR release is completed before tension starts to rise, and other considerations arise. First, it is necessary to suppress the Na/Ca exchange mechanism. Were it not inhibited during the plateau of the action potential, it would rapidly pump out the Ca of the fiber. Second, the known Ca entry via Ca channels appears to be inadequate to produce the level of activation of the contractile machinery that is actually observed. An auxiliary source of Ca is required.

It was suggested in Chapter 4 that changing the membrane potential to the depolarized value found at the level of the plateau allows the entry of Ca via the Na/Ca exchange system running backward.[1] The entry of Ca via both Ca channels and Na/Ca exchange running backward must be reversed during the diastolic interval if the cardiac fiber is to be in the steady state with respect to its internal Ca content. Hence, the Na/Ca system running forward must run rapidly enough to pump out not only the Ca that it introduced itself in running backward but the Ca that entered through the Ca channels as well. The Ca released by the SR must return to the SR and be available for release once again at the next beat.

These considerations are illustrated in Fig. 8.3, which might be appropriate for a tissue such as the cardiac Purkinje fiber. Ca for contraction is equally derived from the three sources that have been mentioned, namely, SR, Ca channels, and Na/Ca exchange. During relaxation, no Ca can flow outward through the Ca channels, since E_{Ca} is more positive than E_m. The SR must take up during relaxation the Ca that it released, and Na/Ca exchange must remove the balance. These sources and sinks for Ca in the fiber are contrasted by the situation for skeletal muscle where, as mentioned above, vir-

---

[1]At the resting potential, the magnitude of I_c is at best a few $\mu A/cm^2$, so that it would require a very long time indeed for this current to depolarize the membrane to the plateau. On the other hand, once I_{Na} has brought the membrane to plateau levels, I_c is capable of influencing the plateau.

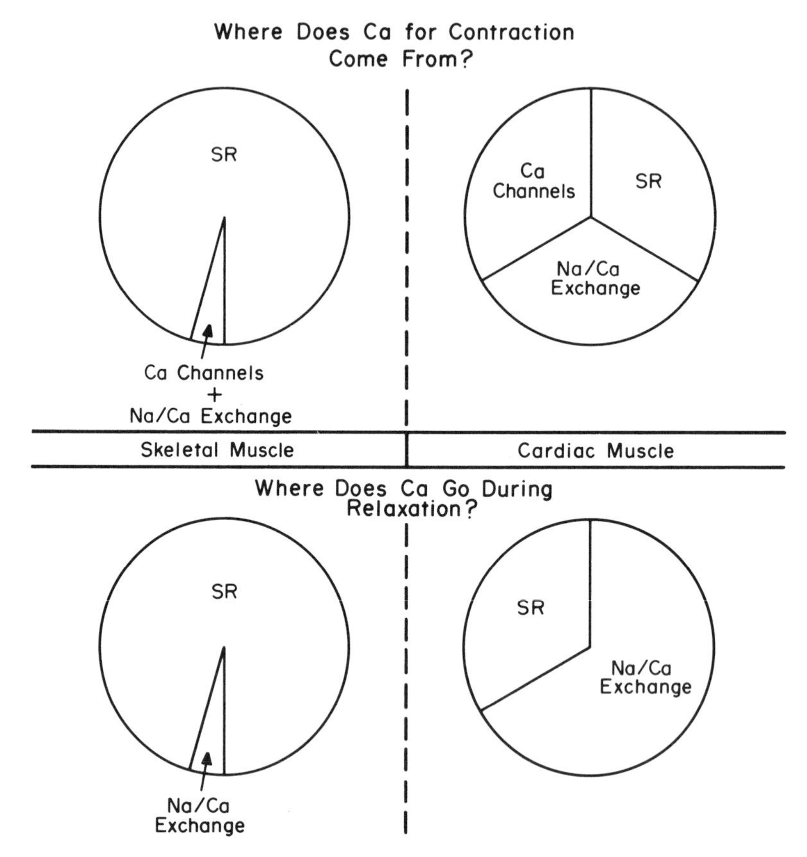

FIG. 8.3. Top: Contrasts the source of contractile Ca for a fast twitch skeletal muscle with that for a cardiac fiber. **Bottom:** The fate of Ca during the period of relaxation for the two fiber types. Note that Na/Ca exchange in skeletal fibers (although normally unimportant) can be made a real source of $[Ca]_o$ by increasing $[Na]_i$ in some fibers.

tually all the Ca for contraction comes from and returns to the SR.

The assignment of an approximately equal quantity of Ca derived from the three sources of Ca that are to be used in contraction to Purkinje fibers is not entirely arbitrary. It is based on a literature that suggests that these fibers may be intermediate between the behavior of frog ventricular fibers, which have scarcely any contribution from the SR, and mammalian ventricular fibers, where Ca derived from the SR may dominate the supply of Ca to the fiber. All cardiac fibers thus far examined by adequate voltage clamp methods have a slow inward current that is at least partly carried by Ca^{2+}. A necessary correlate of this finding is that sufficient Na/Ca exchange pumps must be available to reverse the Ca current during the diastolic interval when the membrane potential is appropriate for an outward movement of Ca. These considerations set a floor below which the density of Na/Ca sites cannot fall. Stated differently, it is the magnitude of the Ca current density and the surface to volume ratio of the fiber that will determine the extent to which it must derive Ca from the outside medium. It is in a large fiber with a small Ca current density and a highly developed SR that the ability of the SR to maximize control over the contractile machinery is achieved.

AMPHIBIAN FIBERS

In frog ventricular fibers, there is evidence that the SR is sparse and that T-tubules are absent. In addition to these anatomical features, which do not favor contributions of Ca from the SR, physiological observations also show that if $[Ca]_o$ is increased after the initiation of a contraction in such fibers, there is an immediate increase in the force of the same beat. If the membrane potential of a frog ventricular fiber is prematurely hyperpolarized, premature relaxation occurs. These observations (4,5) show that contraction is a function of membrane potential and $[Ca]_o$. They do not indicate whether the Ca derived from outside the fiber enters via Ca channels or Na/Ca exchange. Other experiments (6) show that the mechanical response to depolarization increases at potentials beyond (that is, more positive to) E_{Ca}.

These and other observations are difficult to explain except by supposing that at least for extreme depolarizations, all the Ca provided for the contractile response of the fiber is derived from Na/Ca running backward. Under the experimental circumstances outlined, currents through the Ca channels would be outward and thus not deliver Ca to the fiber.

The experiments of Horackova and Vassort (6) illustrate some of the difficulties that have been present in previous investigations. A record of their results is shown in Fig. 8.4 when they change from a normal Ringer's to one in which $[Ca]_o/[Na]_o)^n$ is maintained constant while the voltage clamp applied is to a potential near E_{Ca}, so that a contribution of Ca from I_{Ca} is not possible. The instantaneous effect expected in this experiment is that if the proper exponent for $([Na]_o)^n$ is selected, there will be no initial change in the measured tension in this solution. The steady-state solution is that $[Na]_i$ will decline with time to some new value. Since tension (or at least Ca entry) depends on $[Ca]_o$ $([Na]_i)^n$ the tension will ultimately fall. The record shows that an exponent, n, for the Na concentration of 4 to 5 is one at which the initial tension is maintained; for lower values of the exponent, a rise in tension is followed

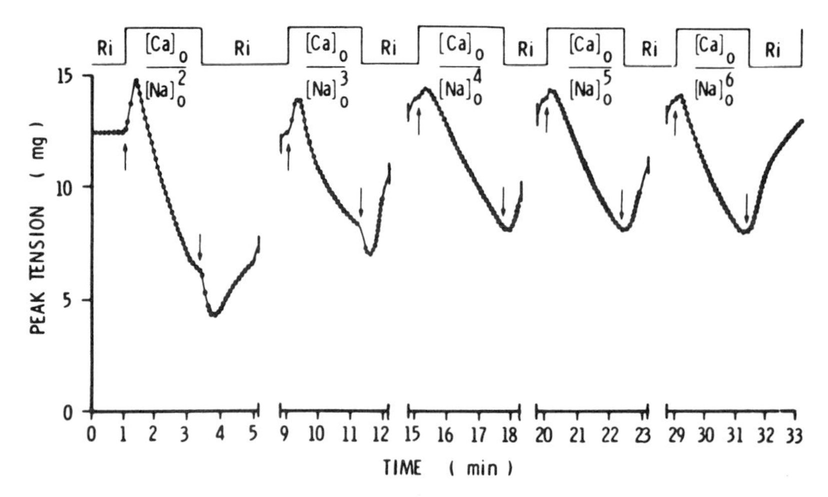

FIG. 8.4. Peak tonic tension elicited by 400 msec and 160 mV depolarizing clamp pulses at a frequency of 15/min. *Arrows,* time of application of normal Ringer's solution (Ri) and solutions with various constant ratios of $[Ca]_o/[Na]_o$ (x = 2, 3, 4, 5, and 6, respectively). In these Ringer's solutions, the $[Ca]_o$ = 10^{-2} mM (buffered with 4 mM EGTA); $[Na]_o$ was (mM) : 8.4, 19.4, 30.8, 39.9, and 46.8 for x = 2 to 6, respectively. (LiCl was used as a replacement for NaCl.) These solutions were buffered with 25 mM Tris to pH 7.2 to 7.4. (From ref. 6.)

by a fall. It is important to work at large values of depolarization. Otherwise, the results may be confused with Ca channel currents. Despite these rather clear results, a massive amount of experimental information suggests that the $[Ca]_o/([Na]_o)^2$ ratio is always obeyed.

Even when experiments are claimed to support a constant $[Ca]_o/([Na]_o)^2$ ratio, the agreement is less than satisfactory. An example is taken from Benninger et al. (7), where data from their Table 1 are plotted in Fig. 8.5. The investigators measured steady tension after a 2-sec voltage clamp in a variety of solutions. The solutions 4 Ca (100% Na) and 1 Ca (50% Na) have a constant $[Ca]/([Na])^2$ ratio and do in fact give equal tension over the entire potential range. However, the solution 0.04 Ca (10% Na), which represents a substantial change in Ca and Na concentrations yet is also a constant ratio, gives vastly smaller tensions for every potential. Clearly, something is wrong. Finally, 0.04 Ca (100% Na) gives substantially less tension than the curve above it, as it should on any basis.

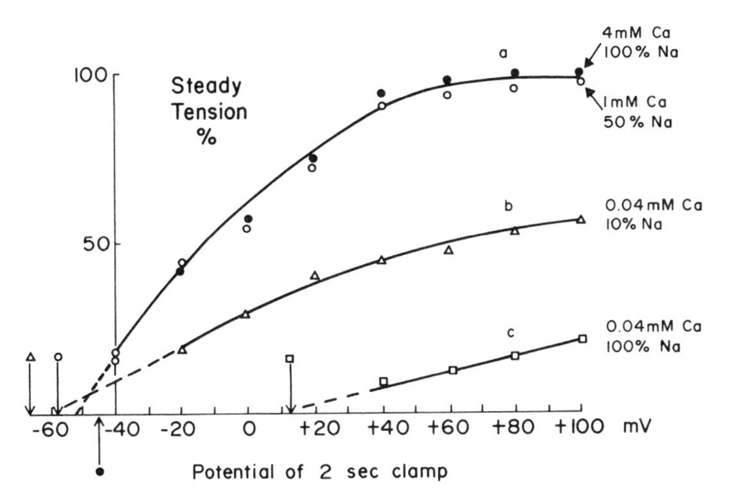

FIG. 8.5. Tension developed in frog atrial fibers over a 2-sec clamp to the potentials indicated and for four different solutions. Curves *a* and *b* are for solutions of constant $[Ca]_o/([Na]_o)^2$ ratio. Along the voltage axis an arrow indicates the calculated E_R of the solution. (Data plotted from Table 1 of ref. 7.)

It is possible to make a rough calculation of E_R for these various experimental situations on the basis that $[Na]_i$ did not change in solutions of altered Na. Thus E_{Na} in 100% Na was 50 mV, and E_{Ca} for $[Ca]_o$ = 4 mM was 145 mV. When this is done, the values of E_R for most of the solutions range between −45 and −67 mV and are probably not different, given the uncertainties about $[Na]_i$ and $[Ca]_i$. The curves do extrapolate at least roughly to zero tension at E_R, even for that value of E_R that is clearly different [0.04 Ca (100% Na)]. A further point is that at a membrane potential of zero (close to the normal plateau), where one might expect maximal Ca release from the SR, there is virtually no tension in low Ca_o high Na_o solution. On the other hand, at +100 mV, where there is virtually no Ca current via Ca channels, there is still a large separation between the three curves, implying that the major source of the Ca is Na/Ca exchange. Finally, at 4 mM Ca versus 0.04 mM Ca (100% Na), the [Ca] differs by 100-fold, but the tension at +100 mV differs by only fivefold. This implies that membrane potential and $[Ca]_o$ sum in their effects on the production of tension.

If the data of Benninger et al. (7) are analyzed at constant $[Na]_o$ but with varying $[Ca]_o$, again from their Table 1, the results are as shown in Fig. 8.6. There is clearly a saturating value of tension that is reached at +100 at a $[Ca]_o$ of 1 mM, while for E = 0, the limiting tension is about 50% of that at +100. These results emphasize that membrane potential is a real variable in the production of tension (along with $[Ca]_o$ and $[Na]_o$), and that membrane potential has effects in a range that cannot plausibly be ascribed to I_{si} or to SR Ca release; i.e., it must have an effect on Na/Ca exchange.

Important conclusions about contractile Ca are presented by Chapman (2), who shows very clearly the following points with respect to the response of frog atria to Na-free solutions: (a) The contractile response is transient, with complete relaxation taking place in a few minutes; complete relaxation can be produced rapidly by replacing Na_o; (b) repriming of the Na-free effect requires time in Na-containing solutions; (c) strophanthidin greatly enhances the sensitivity of the atria to small reductions in $[Na]_o$; and, (d) the tension at reasonably high $[Na]_o$ is given by $T = k\,(1/([Na]_o)^2)^n$,

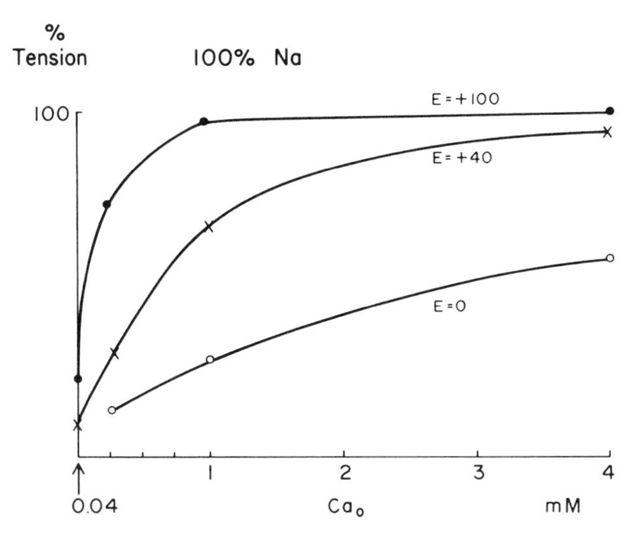

FIG. 8.6. Steady tension 2 sec after a clamp for three membrane potentials as a function of $[Ca]_o$. Tension obviously depends on both $[Ca]_o$ and E_m. (Data plotted from Table 1 of ref. 7.)

with $n = 2$ giving a best fit; i.e., tension varies at the reciprocal of the 4th power of $[Na]_o$.

Figure 8.7 is taken from Fig. 1 of Chapman (2) and shows a number of points that deserve comment. In Fig. 8.7a, a rapid contraction and a slow relaxation are produced by a continued exposure to Na-free conditions. The decline of tension is caused by the decrease of $[Na]_i$. Hence the Ca influx that produced the tension is decreasing, while an internal compartment (presumably the SR) is constantly removing ambient Ca. When Ca entry via a reduction in $[Na]_i$ is less than Ca removal by the SR, the $[Ca]_i$ falls and the fiber slowly relaxes. As proof that Na_i is necessary for this effect, a test with 10 rather than 0.1 mM Ca_o produces no effect. In Fig. 8.7b, the tension trace shows that in a 1 mM Ca_o, the tension production upon making $[Na]_o = 0$ is rapid; upon reapplication of Na_o, the relaxation is equally prompt. That is, the mechanism that delivers Ca and the one that pumps it out are of equal velocity. Fig. 8.7c shows that the absence of the Na_i does not prevent the pro-

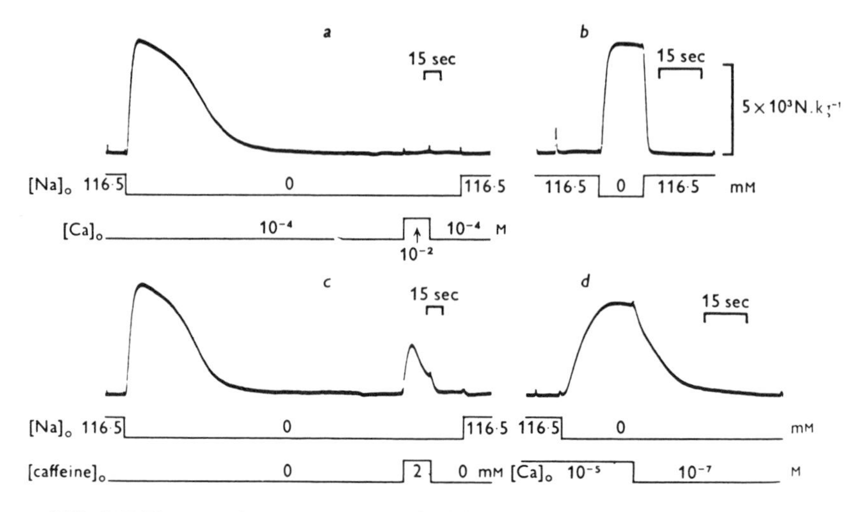

FIG. 8.7. Traces of contractures evoked by perfusion with Na-free solutions. **a:** Na-free contracture shows spontaneous relaxation. If the Ca concentration is raised by as much as 100 times after the spontaneous relaxation, it fails to induce any change in the tension generated by the trabecula. **b:** Contraction initiated by withdrawal of the Na^+ from the bathing solution is rapidly abolished if the Na is returned to the solution (1 mM Ca). **c:** After the spontaneous relaxation of a Na-free contracture, addition of 2 mM caffeine induces a large redevelopment of tension (0.1 mM Ca). **d:** Na-free contracture develops when the $[Ca]_o$ is as low as 10^{-5} M; upon addition of 0.4 mM Tris EGTA (to give a free Ca concentration of less than 10^{-7} M), the trabecula relaxes. This relaxation is slower than that seen when Na^+ are added to the bathing medium. All records are from the same experiment; 20.5° C. The tensions generated by the muscle are expressed in terms of the wet weight of the trabecula; the wet weight of the preparation used was generally between 0.02 and 0.08 mg. (Data taken from ref. 2.)

duction of tension by a caffeine-induced SR release of Ca; Fig. 8.7d shows that when $[Ca]_o$ is made negligibly small, relaxation is slow; it depends not on Ca extrusion by Na/Ca exchange (there is no Na_o), but on the ability of the SR to sequester Ca. This latter process is clearly many times slower than Ca movement via Na/Ca exchange. Points (b) through (d) in the last paragraph clearly support the view that

$$[Ca]_i = \frac{[Ca]_o([Na]_i)^4}{([Na]_o)^4} k$$

An extreme example is the demonstration that a decrease of $[Na]_o$ from normal to 60 mM produces no tension in normal atria, while in strophanthidin-treated fibers, almost maximal tension results.

As a final comment on the proposal that contractility should depend on $[Ca]_o/([Na]_o)^2$, we note that originally (8) it was thought that if the carrier had a negative charge, it would be moved to the inside when the membrane potential was reduced. In a further elaboration of this suggestion, Niedergerke (9) developed equations as follows:

$$[CaR]_i = \alpha([Ca]/[Na]^2) \exp(zFE_m/RT)$$

$$\Delta E_m = (RT/zF) \ln([Na_1]^2/[Na_2]^2)$$

where CaR_i is the delivery of Ca to the inside of the membrane (proportional to contraction), α is constant, and E_m is the membrane potential. For the case where $[Na]_o$ was changed from $[Na_1]$ to $[Na_2]$, the idea was that if $z = -1$ on the carrier, the change in membrane potential and the change in $[Na]_o$ were related as given by the second equation above.

This equation can be rewritten as follows:

$$\Delta E_m - 2E_{Na} = 0$$

Since the equations were developed for a constant tension of $[Ca]_i$ (at constant $[Ca]_o$), they represent

$$E_{Ca} = 2E_{Na} - E_m$$

or

$$2E_{Na} - E_{Ca} - E_m = 0$$

the equation for the equilibrium distribution of Ca given a 4:1 Na/Ca coupling ratio. Confusion between a Na/Ca competition and the reduced equation for the equilibrium distribution of Ca is at the root of the equations that have been developed.

Perhaps the clearest experimental findings that mandate an inward movement of Ca via Na/Ca exchange are those of Anderson et al. (10). Their abstract ends: "The observations suggest that such trans-sarcolemmal calcium movement is brought about by driving forces which add to those of electrodiffusion." The observations are consistent with a Ca entry produced by the Na/Ca exchange carrier running backward at E_{Ca} or beyond where Ca current via I_{si} is either zero or outward.

A final question is whether the curves relating a decrease in either $[Na]_o$ or membrane potential to contraction can be related quantitatively to the calculated properties of the Na/Ca exchange. If we take $[Ca]_o$ = 2 mM, $[Na]_o$ = 140 mM, $[Na]_i$ = 10 or 20 mM, and E_m = −80 mV, then equilibrium $[Ca]_i$ can be calculated from equations given earlier. These yield curves given in Fig. 8.8, where $[Ca]_i$ is plotted as a function of membrane potential for the two values of $[Na]_i$. For comparison, a tension-membrane potential plot from Beeler and Reuter (11) is also given. If it is assumed that contractile saturation occurs around 2 μM Ca_i, then the Na/Ca curves reasonably predict the sensitivity of contraction to E_m. A

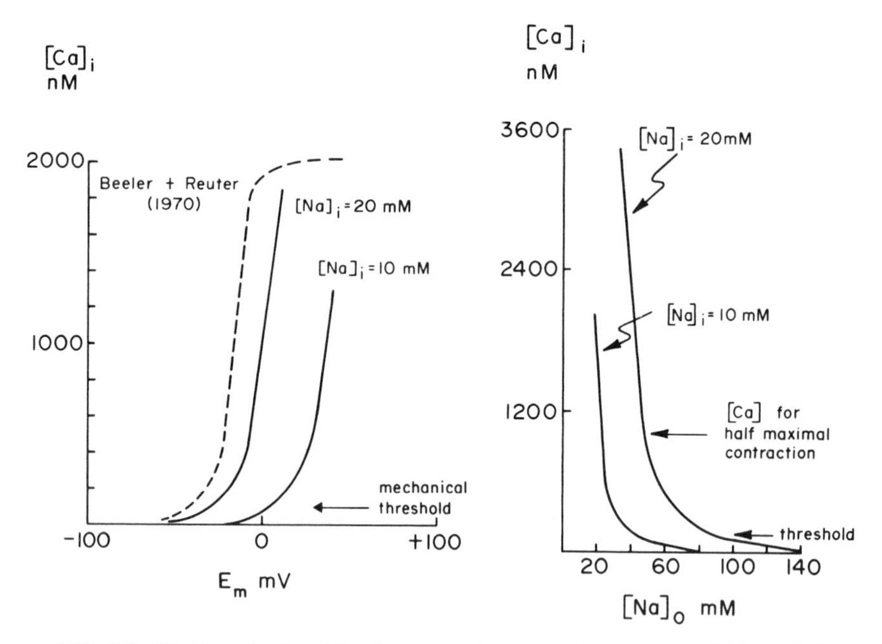

FIG. 8.8. $[Ca]_i$ is calculated for the case where membrane potential is varied **(left)** and where $[Na]_o$ is varied **(right)** according to the relationship $[Ca]_i = [Ca]_o([Na]_i)^4/([Na]_o)^4 \exp(-2EF/RT)$. The arrow labeled mechanical threshold is $[Ca]_i$ = 100 mM; half-maximal contraction is assumed to be $[Ca]_i$ = 1 μM.

similar calculation is made for $[Ca]_i$ versus $[Na]_o$ (instantaneous). These curves show that for $[Na]_i$ = 20 mM, $[Na]_o$ must be reduced to about 45 mM for half-maximal contraction; at $[Na]_i$ = 10 mM, a substantially lower $[Na]_o$ is required.

A direct demonstration that Ca influx with depolarization depends on $[Na]_i$ and is therefore presumably Na/Ca exchange has not been

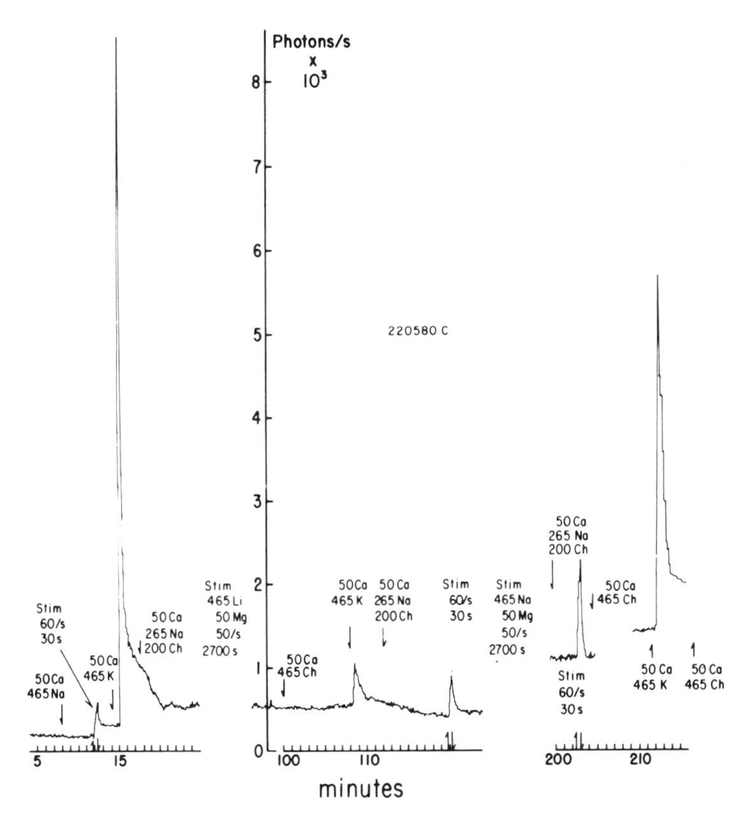

FIG. 8.9. This record of light emission versus time compares an aequorin-injected axon stimulated at 60 per sec for 30 sec with the aequorin response to a steady depolarization from 465 mM K 50 mM Ca seawater. The left trace is for a normal, freshly injected axon; the center trace is the response after the axon has had $[Na]_i$ lowered by stimulation in Li seawater. The responses to 60 per sec stimulation versus steady depolarization differ by 17-fold in an axon with a normal $[Na]_i$; the responses are comparable when $[Na]_i$ is reduced. (From ref. 13.)

obtained until quite recently. Measurements of Ca influx have relied mainly on the measurement of voltage clamp slow inward currents, and these are not suitable for detecting Na/Ca exchange in the presence of Ca channels. In the squid giant axon, however, there is no slow inward current and hence presumably no Ca channels of the sort found in cardiac muscle. In spite of this lack, Baker, Meves, and Ridgway (12) found that in aequorin-injected axons a depolarization gave a large Ca entry; a further study of this effect by Mullins and Requena (13) showed that the Ca entry could be largely abolished by reducing Na_i to half its normal value, or that Ca entry could be very greatly enhanced by raising Na_i 50% above its normal value. A record from their study is shown as Fig. 8.9, where a comparison is made between the response to stimulation of the axon at 60 per sec for 30 sec and the response to a steady depolarization. The former response (which can be shown to be Ca entry via Na channels) is little affected by a change in $[Na]_i$, whereas the response to steady depolarization is reduced 15-fold by a halving of $[Na]_i$. This is thus a demonstration that Na_i-dependent Ca entry is also membrane potential dependent.

REFERENCES

1. Lüttgau, H. C., and Oetliker, H. (1968): *J. Physiol.*, 194:51–70.
2. Chapman, R. A. (1974): *J. Physiol.*, 237:295–313.
3. Jundt, H., Porzig, H., Reuter, H., and Stucki, J. W. (1975): *J. Physiol.*, 246:229–253.
4. Weidmann, S. (1958): *Experientia*, 15:128–129.
5. Kavaler, F., Fisher, V. J., and Stuckey, J. H. (1965): *Bull. NY Acad. Med.*, 41:592–601.
6. Horackova, M., and Vassort, G. (1979): *J. Gen. Physiol.*, 73:403–424.
7. Benninger, C., Einwachter, H. M., Haas, H. G., and Kern, R. (1976); *J. Physiol.*, 259:617–645.
8. Linden, J., and Brocker, G. (1978): *Science*, 199:539–541.
9. Niedergerke, R. (1963): *J. Physiol. (Lond.)*, 167:515–550.
10. Anderson, T. W., Hirsch, C., and Kavaler, F. (1977): *Circ. Res.*, 41:472–480.
11. Beeler, G. W., and Reuter, H. (1970): *J. Physiol. (Lond.)*, 207:211–229.
12. Baker, P. F., Meves, H., and Ridgway, E. G. (1973): *J. Physiol. (Lond.)*, 231:527–548.
13. Mullins, L. J., and Requena, J. (1981): *J. Gen. Physiol. (in press).*

Chapter 9

Mechanisms for Relaxation

INTRODUCTION

The process of relaxation in cardiac muscle is simpler than that involved in the initiation of contraction. Instead of distinguishing three processes that deliver Ca to the myoplasm (Ca via channels, carrier, and SR), there are only two, since the plateau potential is not such that Ca could flow outward via slow channels. There are significant differences in how relaxation is carried out by amphibian and mammalian cardiac fibers. These can be summarized as follows: The amphibian fiber uses Na/Ca exchange to a large extent for relaxation,[1] whereas the mammalian fiber relies to a greater extent on the SR for the initiation of relaxation.

Na/Ca EXCHANGE

During the plateau of the cardiac action potential, it has been suggested that the membrane potential is near E_R; thus there is virtually no carrier current flow. Upon repolarization, a strong driving force is developed that will pump Ca out of the fiber. The rate of Ca extrusion can be expected to depend on $\exp(E_R - E_m)$, and E_R will change as E_{Ca} changes. Relaxation will become complete when $[Ca]_i$ is of the order of 100 nM, but Ca pumping will continue

---

[1]Very clear evidence for a role of $[Na]_o$ in producing relaxation has been provided by Goto et al. (1) for frog atria. This work has been greatly extended by Roulet et al. (2), who were able to show that the time constant for relaxation, which in normal Ringer's solution was 188 msec, increased to 315 msec when Na_o was reduced to 39 mM and to over 2,000 msec at Na_o = 5 mM.

to lower levels of $[Ca]_i$ unless pumping is terminated by a subsequent depolarization (i.e., the next heartbeat).

SR UPTAKE OF Ca

Just as there is little information regarding the way the SR releases Ca in response to a depolarization of the surface membrane, there is virtually no information on whether the SR is able to translate membrane repolarization into a signal for Ca uptake. The following possibilities may be distinguished: The SR Ca pump (a) runs continuously, or (b) stops upon depolarization and starts again upon repolarization. For each of these conditions, we might consider that the SR gating mechanism (a) remains open as long as the fiber is depolarized, or (b) opens only transiently following a depolarization.

In the case of a mammalian fiber with a highly developed SR, when this fiber is subjected to a prolonged depolarization either by voltage clamp or high $[K]_o$, the initial contraction is followed by a relaxation, even though there is no membrane repolarization. This experimental finding would seem to argue that the SR Ca pump runs continuously and that the SR gating mechanism opens only transiently following a depolarization. The SR Ca pump would appear to be running in a depolarized fiber, while Na/Ca exchange is inhibited. Even if the Ca channel gating mechanism were open, $[Ca]_i$ could not decline, since the membrane potential is negative to E_{Ca}. Our hypothesis has, however, other difficulties associated with it. It implies that the initiation of relaxation is brought about by the termination of Ca inflow from both I_{Ca} and I_C. I_C is virtually inactivated early in the plateau, however; thus I_{Ca} is of a magnitude sufficient to hold $[Ca]_i$ at a high level even though the SR is maximally activated to remove the entering Ca.

It has been suggested that it is not depolarization per se but the entering Ca that induces a Ca release. Both in amphibian and in mammalian cardiac fibers, treatments that initiate contraction, such as Na-free or high-K solutions, produce only transient tension responses. In skeletal muscle, a high-K solution also produces a transient tension response. It is generally considered that this is because the SR reaccumulates the Ca it has released although more slowly than if an action potential had been used to trigger the Ca release. Such observations suggest that continual depolarization may make the SR slower in Ca reaccumulation because either gates are incompletely closed or pumping is slowed, or both.

Contractures induced by $[Na]_o$ reduction are in principle simpler to understand since there is virtually no membrane potential change. Ca channels do not open, and the SR cannot receive an electrical signal for contraction. Studies of the relaxation from such contractures in frog atria show that for a short period in a Na-free solution, relaxation upon reapplication of Na_o is rapid, implying that the sole mechanism producing relaxation is the reactivation of Na/Ca exchange. If Na-free conditions are imposed for several minutes and $[Ca]_o \sim 0.1$ mM, a slow, spontaneous relaxation is produced, suggesting that it is the SR or other intracellular Ca complexing mechanisms that are able to bring about the relaxation, but only as $[Na]_i$ falls.

It is possible, in mammalian fibers to induce a large release of internally stored Ca either by CN poisoning (yielding a mitochondrial release) or by caffeine (yielding a SR release) and to prevent tension development by keeping Na_o present. If the external solution is switched to Li or choline Ringer's, tension develops. An analysis of such findings indicates that Ca can be extruded from the fiber faster than its release from internal stores. Thus sarcoplasmic [Ca] does not rise to the threshold for mechanical activation of the contractile machinery. On the other hand, an electrical depolarization of mammalian fibers via a long voltage clamp yields a transient tension response (3,4), suggesting that a repolarizing signal is not necessary for the production of relaxation.

One way of simplifying the analysis of relaxation phenomena is to remove Na from the fiber so that Na/Ca exchange is eliminated. It has been known since 1974 that Purkinje fibers with an elevated $[Ca]_o$ will beat spontaneously in a TEA-substituted Ringer's over periods of many hours. This preparation is difficult to understand on several grounds. First, one would expect that a TEA^+/K^+ exchange would occur as a result of the foreign cation leaking in, and that the membrane potential would collapse. Furthermore, since depolarization in these fibers results from Ca influx via *si* channels, one would expect a buildup of Ca inside the fiber. It is possible that TEA or other outside cations can, at least to some extent, substitute for Na in the Na/Ca exchange; but the problem with a lack of extrusion of the foreign cation remains.

At first it might seem that removing Na from the cardiac fiber would greatly simplify the analysis of the relationship between Ca current and tension, and a study of this sort was carried out by Horackova and Vassort (1976) (5). They found that the time integral of I_{si} and tension were closely related for fibers in Li Ringer's. It should be appreciated, however, that such a fiber (a) must have large quantities of Ca stored, because the Na/Ca pump is running at best weakly in Li solutions; (b) can be expected to undergo a different sort of relaxation cycle, because Ca must now be returned to internal stores and is no longer pumped out of the fiber with each voltage clamp pulse; and (c) must have a higher than normal $[Ca]_i$ if this parameter is related to $[Ca]_{SR}$.

To summarize information about the relaxation process, both Na/Ca exchange and the SR participate in bringing about relaxation, with the relative importance of these two mechanisms varying among different kinds of cardiac cells. No convincing evidence is available that relaxation produced by the SR is voltage dependent. Indeed, it is likely that the rate of SR Ca accumulation depends only on $[Ca]_i$. This would mean that in the relaxed state, if Ca pumping by the SR and Ca leak from the SR were in balance, depolarization would lead to a gated release of Ca from the SR and a Ca entry via Ca channels and Na/Ca exchange. Relaxation would require either the inactivation of Ca entry from outside via both mechanisms or repolarization, which would stop Ca channel entry and reverse Na/Ca exchange. This model sets obvious limits on the rate of SR Ca accumulation: It cannot be many times the rate of Ca entry from outside, or contraction would not occur; it cannot be faster than Ca release from the SR in response to membrane depolarization, or this Ca could never contribute to contractile activation.

BEAT-TO-BEAT EFFECTS

Some of the observed effects of changing the frequency of stimulation of the heart are apparent in the next beat (rather than, for example, a change in $[Ca]_o$, which can affect contraction within the beat). Since the kinetics of both I_{Ca} and I_C should be restored to normal given a 1-sec beat interval, these changes in tension produced in the next beat have been reasonably ascribed to the SR. It has been noted that the expected effect of a long interval without a beat is that $[Na]_i$ in the fiber decreases, since the Na pump continues to run and there is only leakage for a Na load. This change in $[Na]_i$ also reduces $[Ca]_i$ and gives the SR Ca pump a longer time in which to accumulate Ca. Thus a long interval between beats is expected to give the SR a better chance to load and also to deplete myoplasmic Na and Ca concentrations. If the principal supplier of Ca for contraction is Na/Ca exchange, there will be less tension in the first beat; if the principal supplier is the SR, more Ca will be available after a long interval.

An intriguing report (6) has recently appeared in which aequorin was used as an indicator of $[Ca]_i$ and tension and aequorin light were measured in Purkinje fibers. The results illustrate an important point: while the release of Ca from the SR in response to membrane depolarization is relatively fast, the recovery from SR gating inactivation is quite slow. A second conclusion is that given time, the SR continues to accumulate Ca over times of the order of 100 sec. These conclusions are reached from examining Figs. 1 and 2 of Weir (6). If fibers are stimulated at regular intervals, and the interval between stimuli (Fig. 9.1) is made more than 5 sec, little change occurs in the tension developed during an action potential. Light production during a contraction, however, increases continuously as the frequency of stimulation is decreased. The times involved are so long that it is difficult to imagine that either I_{Ca} or I_C

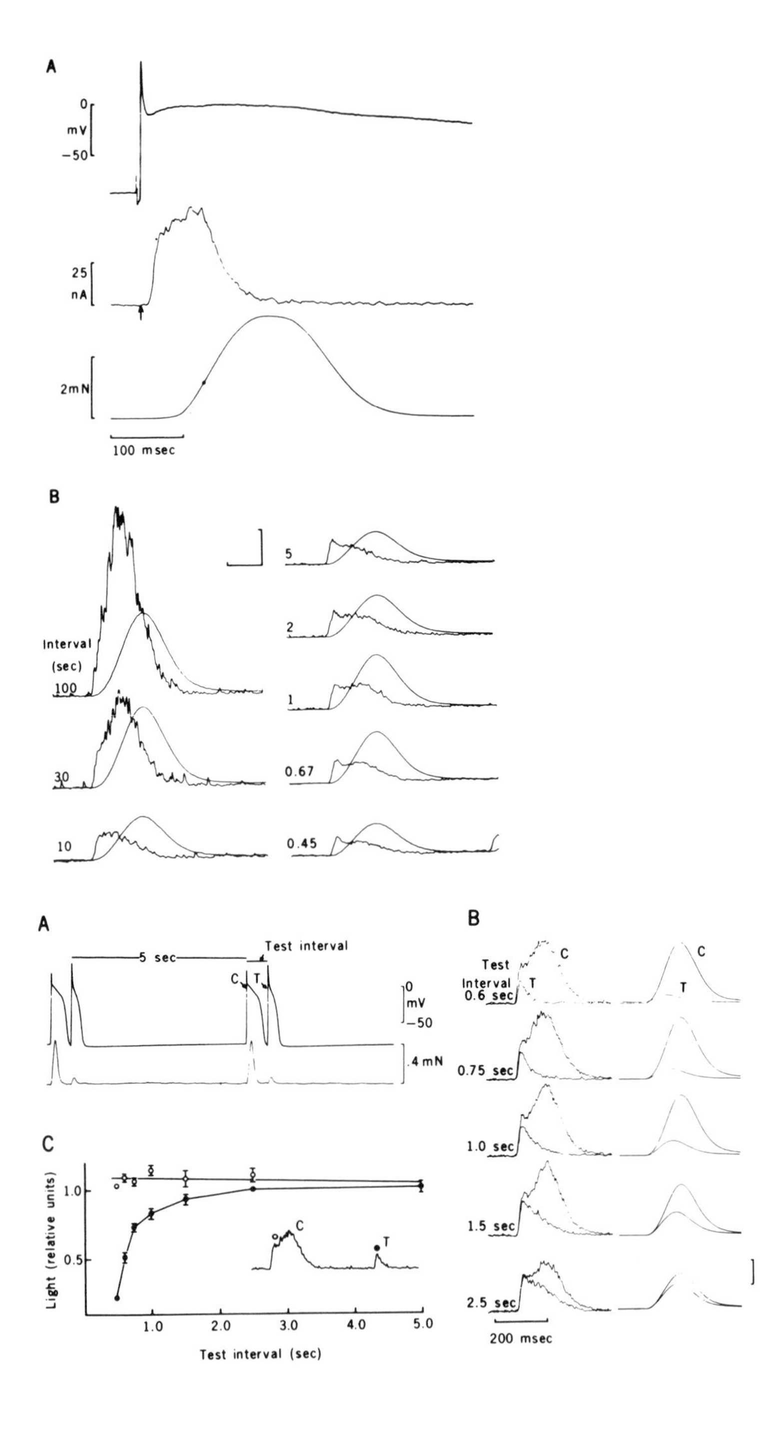
A
0
mV
−50
25
nA
2mN
100 msec
B
Interval
(sec)
100
30
10
5
2
1
0.67
0.45
A
5 sec
Test interval
C
T
0
mV
−50
.4 mN
B
Test
Interval
0.6 sec
0.75 sec
1.0 sec
1.5 sec
2.5 sec
200 msec
C
Light (relative units)
1.0
0.5
1.0
2.0
3.0
4.0
5.0
Test interval (sec)

are changing much with time. It may be supposed that during contraction, with its high $[Ca]_i$, the SR is able to get its main Ca load. Even at the low $[Ca]_i$ of diastole, however, the SR continues to pump Ca from the myoplasm (as does the Na/Ca exchange), so that it continues to load Ca if not interrupted by the need to discharge this Ca following an excitation of the surface membrane.

A second test situation of the experiments of Fig. 9.1 involved using two pulses (a control, first pulse, and test, second pulse) with a variable time interval between them. There was a constant 5 sec time between control pulses. In this experiment, it is clear that two processes are producing the aequorin light that occurs during a pulse. The first is probably that from Ca channels plus Na/Ca exchange, while the second is likely from the SR. If we recall that the responses shown are the result of signal averaging of at least 256 pulses, then it is clear that as the control and test pulses are

←

FIG. 9.1. Top: A, recordings of membrane potential (*top trace*), light (*middle trace*), and force of isometric contraction (*bottom trace*). The arrow marks the time of action potential upstroke as seen in the top trace. The point in the force tracing marks the time at which the rate of force development was greatest. A total of 128 light and force signals were averaged. Paired stimulation: 5-sec interval alternating with 1-sec interval. The responses shown were preceded by a 5-sec interval. Action potential duration was 700 msec. **B,** influence of the interval between stimuli on the aequorin signal and force of contraction. The stimulus interval (in seconds) is indicated at the beginning of each set of superimposed light and force tracings. Vertical calibration bar, 5 nA or 0.5 mN; horizontal calibration bar; 100 msec. **Bottom: A,** illustration of stimulus protocol; records of membrane potential and contractions. Conditioning responses (C) were always preceded by a 5-sec stimulus interval. Test responses (T) were generated after a variable test interval. **B,** light and force records from an experiment in which the protocol was as shown in **A.** The conditioning and test responses at each test interval (indicated) are shown superimposed for the light signals (*left*) and force (*right*). The conditioning response is always the larger one of each pair. Vertical calibration bar, 2.5 nA or 0.4 mN; 128 signals were averaged. **C,** graphic representation of the result of four experiments as shown in **A** and **B.** Points are mean ± SE, except for the single value point at the 0.45-sec interval. *Open circles,* amplitude of initial phase of conditioning aequorin signals; *closed circles,* amplitude of initial phase of test aequorin signals. Amplitudes have been normalized in each experiment to the amplitude of the test signal in that experiment at a test interval of 2.5 sec. Lines were drawn by eye.

placed closer and closer together in time, not only is the first process for the second pulse greatly attenuated, but the second process (presumably the SR Ca release) in the first pulse increases as the second pulse response gets smaller. This finding is understandable on the basis that if pulses are far apart in time, there is a SR release with each of the two pulses (and a need for reaccumulation of Ca); if the pulses are close together in time, the second pulse releases no Ca from the SR, and the SR load is available for the next first pulse to release. This implies that the SR gating process, while it may have a rapid release, inactivates and has a slow recovery from inactivation.

REFERENCES

1. Goto, M., Kimoto, Y., Saito, M., and Wada, T. (1972): *Jap. J. Physiol.*, 22:637–650.
2. Roulet, M. -J., Mongo, K. G., Vassort, G., and Ventura-Clapier, R. (1979): *Pfluegers Arch.*, 379:259–268.
3. Beeler, G. W., and Reuter, H. (1970): *J. Physiol. (Lond.)*, 207:211–229.
4. Morad, M., and Goldman, Y. (1973): *Prog. Biophys. Mol. Biol.*, 27:257–313.
5. Horackova, M., and Vassort, G. (1976): *J. Physiol. (Lond.)*, 259:597–616.
6. Wier, W. G. (1980): *Science*, 207:1083–1087.

Subject Index

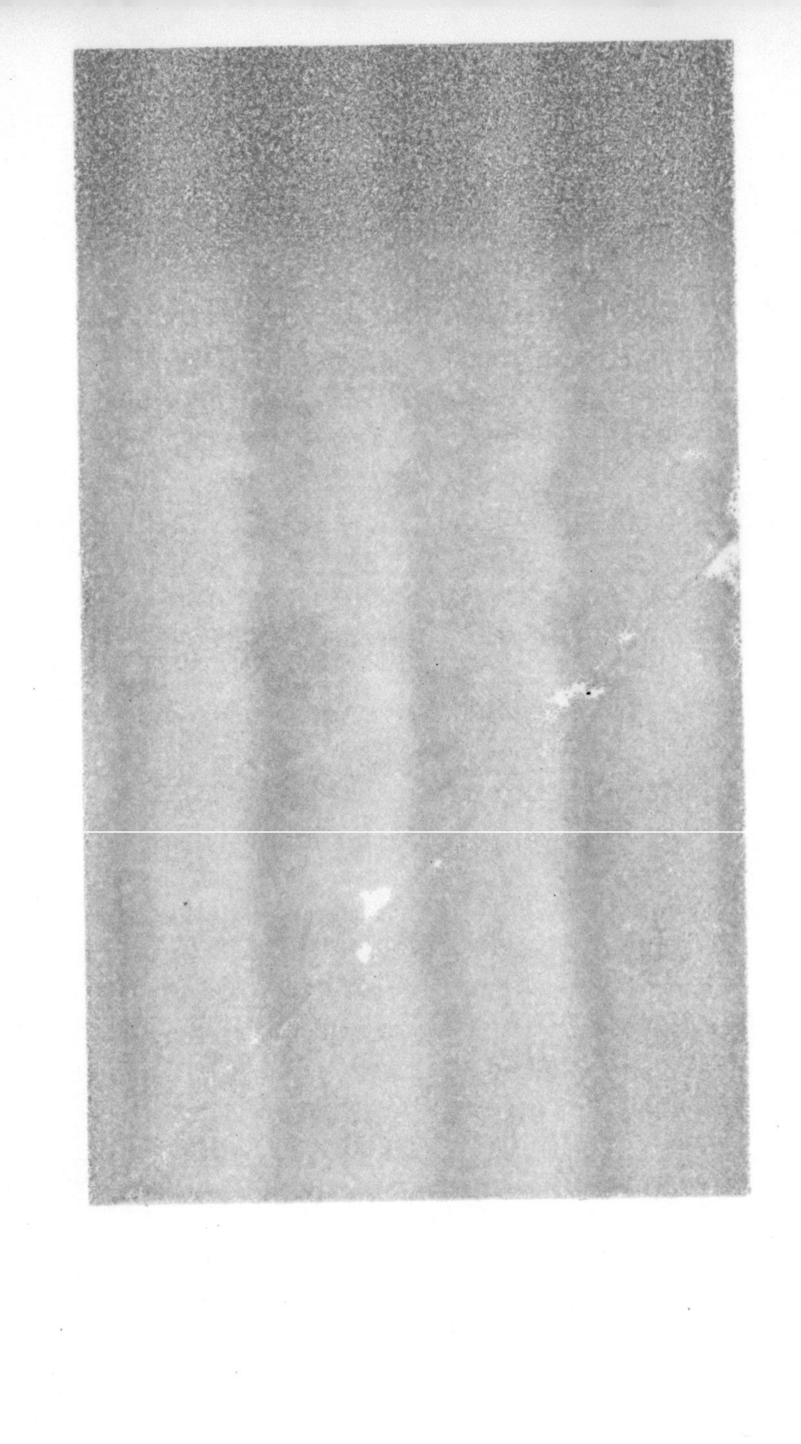